Climate Finance

Theory and Practice

World Scientific Series on the Economics of Climate Change

ISSN: 2010-2127

Published

Vol. 1 The Clean Development Mechanism (CDM): An Early History of Unanticipated Outcomes
by Ariel Dinar (University of California, USA), Donald F. Larson (World Bank, USA) and Shaikh M. Rahman (Texas Tech University, USA)

Vol. 2 Climate Finance: Theory and Practice
edited by Anil Markandya (Basque Centre for Climate Change, Spain), Ibon Galarraga (Basque Centre for Climate Change, Spain) and Dirk Rübbelke (Technische Universität Bergakademie Freiberg, Germany)

World Scientific Series on the Economics of Climate Change
Volume 2

Climate Finance

Theory and Practice

Anil Markandya
Basque Centre for Climate Change, Spain

Ibon Galarraga
Basque Centre for Climate Change, Spain

Dirk Rübbelke
Technische Universität Bergakademie Freiberg, Germany

World Scientific

NEW JERSEY • LONDON • SINGAPORE • BEIJING • SHANGHAI • HONG KONG • TAIPEI • CHENNAI • TOKYO

Published by

World Scientific Publishing Co. Pte. Ltd.
5 Toh Tuck Link, Singapore 596224
USA office: 27 Warren Street, Suite 401-402, Hackensack, NJ 07601
UK office: 57 Shelton Street, Covent Garden, London WC2H 9HE

British Library Cataloguing-in-Publication Data
A catalogue record for this book is available from the British Library.

World Scientific Series on the Economics of Climate Change — Vol. 2
CLIMATE FINANCE
Theory and Practice

ISBN 978-981-4641-80-7

Desk Editor: Jiang Yulin

Typeset by Stallion Press
Email: enquiries@stallionpress.com

Contents

Chapter 1

Challenges in International Climate Finance

Ibon Galarraga, Anil Markandya and Dirk Rübbelke

As long ago as 2010, at the 16th Conference of Parties (COP) to the United Nations Framework Convention on Climate Change (UNFCCC) in Cancun, developed countries formally pledged to mobilize USD 100 billion in international climate financing annually by 2020. These funds were to be channelled towards both mitigation and adaptation projects in developing countries.

Five years later, at the COP 21 in Paris in 2015, it was further agreed that "developed countries intend to continue their existing collective mobilization goal through 2025 in the context of meaningful mitigation actions and transparency on implementation" and, prior to 2025, "a new collective quantified goal" shall be stipulated. The new collective goal must not undercut the 2010 commitment to provide USD 100 billion annually.

A lot of controversy still prevails concerning the more concrete subsequent policy steps, although the Paris deal involves a "Decision" text to give effect to the main "Paris Agreement". There is, for example, still the absence of a clear internationally agreed definition of what counts as climate finance. Furthermore, it is not clear how much individual countries have to contribute, and when.

The disbursement scheme also remains imprecise, for example, concerning the allocation of funds for adaptation projects and mitigation projects. While the latter constitute global public goods [Kaul *et al.*, 1999], this is rarely the case for the former. Although the need for resources for adaptation is recognized in the Paris Agreement, there is no quantified funding target for that component. Yet, one may argue that, as stated in the earlier Cancun Agreement, adaptation "must be addressed with the same priority as mitigation" [UNFCCC, 2011].[1] Thus, it could be assumed that a large share of the USD 100 billion pledge should be dedicated to adaptation support. This perspective could — in turn — be challenged; for example as Schelling [2012] points out: "Eventually (decades from now) aid to the poor may be necessary for their successful adaptation to a changing climate. But the more urgent priority is to reduce carbon emissions where they are rising apace with development." Thus, there is controversy not only concerning the distribution of funds between adaptation and mitigation projects,[2] but also concerning the right timing of specific support alternatives.

These and related controversial aspects associated with climate finance are the focus of the different contributions included in this book. It provides ideas for, and presents research on, a range of approaches to raise, manage and allocate international funds in the climate change context.

The book includes 14 chapters, divided in three parts. The first covers practical issues relating to climate finance: who are the main players, what do they contribute, how political factors affect the allocation of funds and how climate finance structure is evolving? The second provides a more theoretical perspective. Climate

[1]A recent study by Buchner *et al.* [2011] found a split between mitigation and adaptation finance of 95:5. The causes of this mitigation bias in international climate finance has been analyzed by Abadie *et al.* [2013].

[2]There is a growing strand of literature analyzing international transfers taking into account both mitigation and adaptation options, e.g., Buob and Stephan [2013], Schenker and Stephan [2014], Heuson *et al.* [2015] and Pittel and Rübbelke [2015].

finance supports the provision of a global public good (reduction in greenhouse gases), as well as one that is mostly national (adaptation to climate change). The rules for allocation of funds will influence how much of each kind of public good is provided. It is not always the case that the present approaches turn out to be Pareto efficient — i.e., they do not provide a combination of the two goods that cannot be dominated by one in each no one is worse off and some countries are better off. Furthermore, the present system would not be considered as the most equitable by those concerned with the ethical dimensions of the problem. This section looks at the extent of current deviations from an efficient solution and ways in which it could be made more efficient and more equitable. The third part of the book looks at particular tools for appraising investments in climate finance so as to select the options that give the highest combination of benefits in a context of uncertainty.

After this introduction, four chapters are devoted to offering an overview of the current practice in climate finance. The first, by **Román, Ansuategi and Markandya** (Chapter 2) offers an overall review of the state of international climate finance in a very detailed assessment on what is happening in the international arena.

Chapter 3 by **Trabacchi and Buchner** focuses on financing adaptation in the context of the post-Paris climate policy. They acknowledge the difficulties in gathering reliable data and propose improvements in current tracking systems. Their study estimates that during 2013 and 2014 close to USD 25 million had been invested in public adaptation in developing countries.

How the private sector may be engaged in mobilizing climate finance is analyzed in the contribution by **Averchenkova** in Chapter 4. The author argues that low carbon development strategies, Nationally Appropriate Mitigation Actions (NAMAs) and, in the future, the Intended Nationally Determined Contributions (INDCs), will play a central role in redirecting private investment towards climate finance. The public sector can effectively contribute to this with policies to reducing risk, by lowering the cost of capital, enhancing the competitiveness of low-carbon technologies, and building capacity for public-private collaboration in climate finance.

Clapp and Pillay devote Chapter 5 to illustrating the functioning and the role of Green Bonds in the context of climate finance. The Green Bond market is growing fast mainly as a result of the actions of multilateral development banks and, lately, also of corporations and municipalities issuing their own bonds. They discuss the possible application of green bonds to support developing country actions on climate, and present some challenges and opportunities that may arise.

Chapter 6 by **Brumme** looks at the role of critical natural resources in the cost of mitigation technologies. As some of the turbines used in wind energy generation require scarce resources such as rare earth elements, the expansion of wind power technologies may be compromised if the distortions in the markets for these are not reduced. The author discusses measures for making these markets more efficient.

The second part of the book deals with understanding the theoretical basis of climate finance, their impacts and how incentives may work. The first contribution by **Kaul** in Chapter 7 considers climate finance through the lens of the concept of global public goods.

Chapter 8 by **Schenker and Stephan** looks at the role of adaptation funding and shows that positive but smaller than proposed transfers in climate finance may be desirable to achieve Pareto improving situations under the public good character of adaptation funding.

In Chapter 9, **Peters, Schwarze and Topp** look at the post-Kyoto international climate finance regime, unravelling how voluntary sustained finance can be maintained over time overcoming the so-called Bergstrom paradox (i.e., a situation where receiving countries end up worse off as a consequence of conditional climate finance). They find that only those resources devoted to mitigation activities can increase the global efforts in that direction and induce Pareto improvements.

Buchholz, Cornes, Peters and Rübbelke also deal with the Bergstrom paradox in international climate finance in Chapter 10. They show that restricting the scope for offsetting behavior is an effective way to improving the effectiveness of conditional transfers as an instrument of international climate protection. As the authors argue, the mechanism as proposed by them can also be understood as a compromise between the donor and the recipient countries searching for a solution which splits their gains in a fair way.

Pittel and **Rübbelke** in Chapter 11 look at the effects of international transfers on global mitigation and at the repercussions that these transfers might have on welfare in the transfer-providing and the transfer-receiving countries. In their analysis, the authors distinguish between conditional transfers provided in monetary terms (subsidizing mitigation efforts) and conditional in-kind transfers in the shape of support in adaptation to climate change.

Chapter 12 by **Buchholz, Dippl and Eichenseer** deals with the role of technological transfer in climate policy, which — as an indirect instrument for supporting abatement activities in less-developed countries — serves as a substitute for climate finance in the narrow sense. The authors show that leading behavior by some countries based on green innovation together with a strategy of costless technology transfer proves to be a very effective way of fighting climate change, in particular since only the prospect that other countries make use of the improved technology may create an incentive to engage in "green" Research and Development (R&D).

The final part of the book is composed of two chapters devoted to illustrating the use of applied methodologies to assess climate finance related investments.

The contribution by **Chevallier** and **Goutte** in Chapter 13 illustrates the interactions between the electricity and carbon markets using stochastic models for risk management. These are techniques used in the area of financial econometrics. Their findings show that jump models fit better to explain these risks and help set the right incentives for mitigation in complex carbon and electricity markets such as those in Europe and North America.

In Chapter 14, **Abadie** and **Galarraga** use stochastic modelling for carbon and electricity prices, to illustrate the case of real option techniques to adequately assess irreversible investments under uncertainty and investment in projects with a risk of default. The chapter shows that uncertainty plays a very important role in investment valuation especially in the case of climate finance investment.

Climate finance is the life blood of actions to reduce future climate impacts and to adapt to those impacts that will occur. This book provides a guide to the main areas of debate on the topic from a political, institutional and economic perspective. It is by no means the last word on the topic, which is fast evolving, but it does lay out the state of the art on the subject.

Acknowledgments: We thank Mavi Román (BC3) and Claudia Baldauf for their excellent assistance work. We also acknowledge the financial support of the Research Council of Norway under the project "Strategic Challenges in International Climate and Energy Policy (CICEP)" (http://www.cicep.uio.no/english/).

References

Abadie, L. M., I. Galarraga, and D. Rübbelke. 2013. An Analysis of the Causes of the Mitigation Bias in International Climate Finance. *Mitigation and Adaptation Strategies for Global Change* 18(7), 943–955.

Buchner, B., A. Falconer, M. Hervé-Mignucci, C. Trabacchi, and M. Brinkman. 2011. *The Landscape of Climate Finance.* Climate Policy Initiative (CPI) Report. Venice, Italy.

Buob, S. and G. Stephan. 2013. On the Incentive Compatibility of Funding Adaptation. *Climate Change Economics* 4(2).

Heuson, C., W. Peters, R. Schwarze, and A.-K. Topp. 2015. Voluntary International Climate Finance under the Post-Kyoto Framework: The Strategic Consequences of Different Modes of Funding. *Climate Change Economics* 6(3).

Kaul, I., I. Grunberg, and M. A. Stern. 1999. Global Public Goods: International Cooperation in the 21st Century. New York, NY and Oxford, UK: Oxford University Press.

Pittel, K. and D. Rübbelke. 2015. On Fostering International Public Good Provision: Would Complementarity between Public Good and In-kind Transfers Help?. *Economics Bulletin* 35(3), 1638–1644.

Schelling, T. 2012. *Norms, Conventions, and Institutions to Cope With Climate Change.* In: Climate Change and Common Sense — Essays in Honour of Tom Schelling, R. W. Hahn and A. Ulph (eds.), Oxford, UK: Oxford University Press, pp. 19–28.

Schenker, O. and G. Stephan. 2014. Give and Take: How the Funding of Adaptation to Climate Change Can Improve the Donor's Terms-of-trade. *Ecological Economics* 106, 44–55.

United Nations Framework Convention on Climate Change (UNFCCC). 2011. *The Cancun Agreements: Outcome of the Work of the Ad Hoc Working Group on Long-Term Cooperative Action under the Convention.* FCCC/CP/2010/7/ Add.1. At: http://unfccc.int/resource/docs/2010/cop16/eng/07a01.pdf.

Part 1

Practical Perspective

Chapter 2

Climate Finance and International Negotiations on Climate Policy

María V. Román

Basque Centre for Climate Change, BC3,
Sede Building 1-1, Scientific Campus of the
University of the Basque Country, Barrio Sarriena s/n, 48940
Leioa, Spain
Mavi.roman@bc3research.org

Alberto Ansuategi

Foundations of Economic Analysis Department,
University of the Basque Country,
Avenida Lehendakari Aguirre 83, 48015
Bilbao, Spain
Alberto.ansuategi@ehu.es

Anil Markandya

Basque Centre for Climate Change, BC3,
Sede Building 1-1, Scientific Campus of the
University of the Basque Country, Barrio Sarriena s/n, 48940
Leioa, Spain
Anil.markandya@bc3research.org

Abstract

The Paris Agreement adopted in December 2015 includes climate finance as a key element of the coordinated action against climate change at global level. This chapter deals with the role of transboundary climate finance in the international negotiations on climate change. After discussing the magnitude of the challenge, and synthesizing the insights from the literature on international environmental agreements, we describe the complexity of the current architecture and identify the main opportunities and challenges of the post-Paris regime.

1. Introduction

Climate change poses enormous financial challenges to the international community. The combined cost of mitigating and adapting to climate change has been estimated at several hundred billion US dollars per year [OECD/IEA, 2015; UNEP, 2014]. Furthermore, a large share of these financial resources will have to be invested in developing countries, which do not have easy access to the required finance. Thus, international climate finance is one of the key challenges for global climate action.

The term "climate finance" within the academic literature [Glemarec, 2011; van Melle *et al.*, 2011] refers to climate-related financial flows within or between countries that are dedicated to both mitigation and adaptation.[1] In this chapter, we will focus exclusively on the international flows of funds, and more precisely in those flows from developed to developing nations to help poorer countries cut their emissions and adapt to climate change.

Under the United Nations Framework Convention on Climate Change (UNFCCC), developed countries committed to support developing countries with new and additional financial resources to cover full incremental costs of agreed mitigation measures, the cost of adaptation and the cost of the preparation of their national communications. Indeed, the agreements indicate that the extent to which developing countries will implement their commitments under the UNFCCC depends on the effective implementation of these financial

[1] Although clear in principle, it is very difficult to classify some expenditures as adaptation or mitigation, especially when they involve investments with multiple objectives. Further work on how to measure and classify expenditures more accurately is needed.

commitments, together with the transfer of technology [Gichira *et al.*, 2014].

In the Copenhagen Accord (in 2009), developed countries promised Fast-start Finance (FSF) of about US$10 billion per year from 2010 to 2012. This Accord also included a commitment by developed countries to jointly mobilize US$100 billion per year by 2020 to address the needs of developing countries, which had been recently reaffirmed in the Paris Agreement in 2015 [COP21, 2015]. The Conference of Parties (COP) had further agreed that funds may come from a variety of sources, including public, private, bilateral and multilateral alternative ones [Edenhofer *et al.*, 2014].

This chapter aims at shedding some light on the role of climate finance in international negotiations on climate policy and is structured in five sections. After this brief introduction, Section 2 will show that the magnitude of climate finance required for developing countries will be determined by both climate protection objectives and the burden-sharing rules adopted. Section 3 will discuss the role of North-to-South financial transfers as a way to provide economic incentives to regions to participate in international climate agreements. In Section 4, we will try to understand the current and future landscape of the climate finance architecture in the light of international climate negotiations. Finally, Section 5 will draw some conclusions.

2. Equity Principles, Burden-Sharing Rules and Climate Finance

In the context of climate change, considerations of equity concern the relations between individuals and groups of individuals (countries) both at a single point in time (intra-generational equity) and across time (intergenerational equity). In this section, we will adopt the perspective of countries rather than individuals and will primarily focus on intra-generational equity aspects.

Climate change raises important ethical questions at the international arena such as how much overall climate mitigation is needed to avoid "dangerous interference", how the effort or cost of mitigating climate change should be shared among countries, how to account for factors such as historical responsibility for emissions, and how to choose among alternative policies for mitigation and adaptation.

Article 3.1 of the UNFCCC states: "*The Parties should protect the climate system for the benefit of present and future generations of humankind, on the basis of equity and in accordance with their common but differentiated responsibilities and respective capabilities.*"

Both the UNFCCC and the Kyoto Protocol, when dealing with the general fairness norms that should guide global cooperation distinguish explicitly between developed and developing countries (North and South, respectively, hereafter). Here we will also center on finance flows channeled from North to South for climate change mitigation and adaptation projects and programs.

Financial flows directed from North to South are partially motivated by reasons of equity and fairness [Grasso, 2010] but economic and political reasons also play an important role within international negotiations [Abadie *et al.*, 2013]. The magnitude of climate finance flows from developed to developing countries is highly uncertain mainly due to uncertainty about the scale of the private flows. One of the first approximations of the climate finance flows from developed to developing countries presented a wide range of US$40 to US$175 billion for the period 2010–2012 [UNFCCC, 2014]. The lower bound of that range was taken from Buchner *et al.* [2014], who provided an average estimate of US$34 billion for that period, 94% coming from the public sector (US$32 billion). Stadelmann *et al.* [2013], who tackle the task of tracking private flows, provided the upper bound. More recently the Organisation for Economic Co-operation and Development (OECD), in collaboration with Climate Policy Initiative (CPI) (a leading think tank on the subject), presented preliminary and partial estimates for the period 2013–2014: US$57 billion per year on average, split into US$40.7 billion from public sources (71% of the total), US$1.6 billion from export credits (3%) and US$14.7 billion from private sources (26%) [OECD, 2015a].[2]

[2]There is considerable uncertainty about these figures and it is important to note that the developed to developing country flows are only a small part of total

Ringius *et al.* [2002] explore which basic principles of fairness are sufficiently widely recognized to serve as a normative basis for a scheme of "burden-sharing"[3] that can be accepted as "fair" by all or at least most governments taking part in global climate change negotiations. They find that the principles of "responsibility", "capacity" and "need" are the three main principles of equity invoked by important groups of actors. Consequently, a burden-sharing rule aimed at being widely accepted should somehow preferably combine all three of these principles (allowing for exemptions).

Höhne *et al.* [2014] compare over 40 studies that analyze future greenhouse gas (GHG) emissions allowances or reduction targets for different regions based on a wide range of effort-sharing approaches and long-term concentration stabilization levels. They conclude that distributional impacts differ significantly with underlying criteria for effort-sharing and that the stabilization level and the shape of the global emissions pathway have a significant influence on the results for all regions. For instance, for the stabilization level required for having a medium chance of reaching the 2°C target, emissions changes from 2010 levels in 2030 range from −37% to −75% in OECD 1990 countries, −28% to −53% in Eastern Europe and Russia, +7% to −33% in Asia, +24% to −7% in Middle East and Africa, and −15% to −49% in Latin America, depending on the effort-sharing approach.

Bowen *et al.* [2015] employ a number of Integrated Assessment Models (IAMs) to determine what financial transfers would have to take place between developed and developing economies if the climate mitigation effort, measured as mitigation costs as a share of gross domestic product (GDP), were to be equalized across regions. They find that in four out of six models a financial transfer of around

developing country climate finance. The CPI estimates that from the US$165 billion spent in developing countries on climate related mitigation and adaptation in 2013, around 80% came from domestic or other non-developed country sources [Buchner *et al.*, 2014].

[3] "Burden-sharing" rules specify how to share the cost of the provision of collective goods or common initiatives between states. In the context of climate change, it usually refers to sharing the cost of mitigation, although it could also be applied to international support for adaptation.

US$400 billion per year is required from developed to developing countries by 2050, while the other two models imply even larger sums (US$2,000 billion a year).

Most analyses of equity principles and burden-sharing focus on mitigation. Ciplet *et al.* [2013] discuss some likely reasons for the low profile of adaptation in the early stages of the debate and go on to explain how adaptation finance has recently risen to the top of the UNFCCC negotiations. Ringius *et al.* [2002] argue that issues of equity and burden-sharing should be within a consistent framework that integrates the total costs and benefits of mitigation and adaptation, as well as the costs of local climate damages. The United Nations (UN) estimates of annual adaptation costs for developing countries span the range from US$17 billion to US$100 billion [UN/DESA, 2009]. A World Bank study puts this costs at between US$70–100 billion a year [World Bank, 2010]. A more recent work on the adaptation gap has tried to estimate the costs as a function of different climate scenarios and suggests there is little difference in adaptation expenditures between the 2°C and 3–5°C scenarios up to 2030 [Watkiss *et al.*, 2014]. However, by 2050 a low ambition agreement has very fast rising adaptation costs, and is estimated to cost around US$66 billion per year, compared to US$46 billion per year that is the estimated adaptation cost of a high ambition agreement. According to the ethical principles of historical responsibility and capacity to pay, Annex I countries would have to provide funding to developing countries for up to 70% of their needs (between US$65 and US$70 billion per year, according to Dellink *et al.*, 2009).

Table 1 presents estimates of financial transfers resulting from burden-sharing principles together with estimates of the actual climate finance flows gathered from the literature. The main point to note is that all the estimates of needs (in the range of US$400–2,100 billion per year) are above the amount currently provided from international sources (US$40–175 billion per year). Therefore, a lot depends on future increases of adaptation funds, along with the application of mitigation measures that are consistent with sustainable development priorities (i.e., better environmental policies, carbon

Table 1. Estimated Needs vs. Actual Finance Flows.

Reference	Theme	Geographies	Range (USD billion/year)	Time period	Type of estimation
Tian and Whalley, 2010	Mitigation	EU, US & Japan to the BRICS	[1,143–1,482]	2006–2056	Transfers needed to compensate utility losses from taking on a 50% emission reduction target from the Business as Usual (BAU) scenario estimated with a General Equilibrium Model
Bowen *et al.*, 2015	Mitigation	North to South	[400–2,000]	2050	Transfers mobilized by a theoretical carbon market if climate mitigation effort were to be divided equally across regions estimated with several Integrated Assessment Models
UN/DESA, 2009	Mitigation/ Adaptation	Developing countries	[250–950]/ [16.6–97.47]	2030	Additional investments needed for the +2°C scenario based on a review
World Bank, 2010	Adaptation	Developing countries	[70–100]	2010–2050	Needs of adaptation finance in the +2°C scenario based on a review of national databases
Dellink *et al.*, 2009	Adaptation	North to South	[65–70]	2015	Transfers resulting from the distribution of US$100 billion of adaptation costs according to the responsibility and capability principle

(*Continued*)

Table 1. (*Continued*)

Reference	Theme	Geographies	Range (USD billion/ year)	Time period	Type of estimation
Watkiss *et al.*, 2014	Adaptation	Developing countries	[46–66]	2050	Adaptation expenditures needed in the 2°C and 3–5°C scenarios
UNFCCC, 2014	Mitigation and adaptation	North to South	[35–50] [5–125]	2010–2012	Actual public flows based on a review/Actual private flows based on a review
Buchner *et al.*, 2014	Mitigation and adaptation	North to South	[31–37]	2013	Actual public flows based on primary data of public sources
Stadelmann *et al.*, 2013	Mitigation and adaptation	North to South	[27–123]	2008–2010	Actual private flows based on primary data of private sources
Edenhofer *et al.*, 2014	Mitigation and adaptation	North to South	[10–72]	2008–2011	Actual private flows based on a review
OECD, 2015	Mitigation and adaptation	North to South	[37.9–43.5]/ 1.6/ [12.8–16.7]	2013–2014	Actual public/export credits/private flows based on a review

finance and energy efficiency). Another relevant point to note is that, although funding needs for mitigation are much larger than those of adaptation, mitigation investments generate savings in fossil fuels expenditures that offset the initial investment costs [OECD/IEA, 2015].[4] These net benefits of mitigation, together with the inverse relationship between mitigation ambition and adaptation costs, create a compelling case for supporting mitigation. Besides, increasing the adaptation funding is necessary to increase resilience to climate change and reduce its impacts on the poorest and most vulnerable communities. Adaptation funds will address the increased vulnerability associated with climate change, especially to ensure that those who may fall into poverty or whose level of poverty may increase, are prioritized in terms of assistance. Otherwise, there could be an impact on poor and vulnerable communities and this would be much larger after 2030 for the low ambition climate agreement.

3. The Role of Climate Finance in International Climate Negotiations

The previous section provided an overview of the literature dealing with the transfers associated with different effort-sharing schemes to achieve a given set of mitigation and adaptation targets. But optimal transfers can also be devised as a way to provide incentives to regions to participate in international climate agreements. The theory of coalition formation shows that international transfers are a fundamental element to achieve a broad participation in an environmental international agreement. International transfers attract participation because they balance the asymmetric gains of cooperation. Transfers can be monetary, technology or indirect transfers via market-based mechanisms [Edenhofer *et al.*, 2014]. Here we will mainly focus on monetary and indirect transfers.

In an international agreement on climate change mitigation, participation of major emitters is essential for the agreement to be

[4]The Energy Technology Perspective model estimates the annual net savings of mitigation for the +2°C scenario in more than US$2,200 billion per year globally.

effective. Otherwise, carbon leakage would undermine the achievements of the participants in the agreement [Barrett, 2010; de Zeeuw, 2015]. Greater participation not only improves the environmental effectiveness, but also the economic performance of the agreement because it enables lower-cost mitigation options [Edenhofer *et al.* 2014].

Tian and Whalley [2010] confirm that climate finance helps in attracting main emitters in the developing world to the agreement and that the volume of transfers required depends on the ambition of the agreement. If the agreement is very demanding with this group of countries, i.e., requiring from them large reductions in emissions, then the transfers that would persuade them to join in are also large. As the mitigation burden corresponding to these countries decreases, so does the climate finance flows required by them to participate. For example, if the so-called BRICS (Brazil, Russia, India, China and South Africa) had to reduce their emissions up to 50% with respect to the baseline scenario, the climate finance required would be between 3–5% of their GDP (US$1,143–1,482 billion per year).

Apart from making participation more attractive from an economic point of view, climate finance also influences another aspect of the agreement that is a key element for participation: fairness. Theories of justice provide different ways of conceptualizing fairness. According to Schmidtz [2006], such theories are typically assembled from one or more of the following four principles: equality (people/countries should get equal shares of whatever is distributed), need (those who are worse-off should be paid special attention), desert (people/countries ought to get what they deserve), and reciprocity (people/countries' positive actions are rewarded and harmful actions are punished).

Climate finance may contribute to "distributive fairness" (encompassing both the equality and need principles), leading to a more balanced distribution of costs and benefits between countries. This is important, because countries would hardly be willing to participate in an agreement if they consider it unfair from a distribution perspective [Bodansky and Diringer, 2014; Edenhofer *et al.*, 2014; Pickering *et al.*, 2015].

But an egalitarian distribution of costs and benefits is not the only factor that might influence the perception about how fair an agreement is. Pittel and Rübbelke [2013] suggest two variants of what could be "reciprocal fairness" namely: (a) "intention-driven fairness" (dependent on beliefs of other countries' intentions and therefore incorporating the desert principle); and (b) "consequence-driven fairness" (based on the consequences of activities executed by other countries before the present negotiation takes place and therefore incorporating the reciprocity principle). The authors argue that, due to these two considerations, climate finance will influence the willingness of recipient countries to participate. For example, support for mitigation can be seen as a sign of predisposition to participate by donors, stimulating a (reciprocal) positive response from recipient countries that may end up taking part. Support for adaptation could also mitigate the unfairness associated with the participation of non-responsible countries, if it is seen as reparation for the damages already caused by donors. On the contrary, if developed countries fail to comply with their promise of providing financial support, this could negatively affect trust in developed countries' serious commitment, and trigger a (reciprocal) negative response from developing countries. Pickering *et al.* [2015] also remark that climate finance demonstrates good faith on the part of developed countries, whereas failing to meet this promise would damage the credibility of the UNFCCC regime.

The experience of previous international environmental agreements such as the Montreal Protocol shows how North-to-South transfers can pave the way for major developing countries to join the agreement [Edenhofer *et al.*, 2014]. In the case of that Protocol, international financial support helped to create the required capacities in recipient countries and an environment (including regulation, national development strategies, economic instruments, etc.) compatible with the compliance of the agreement's obligations. Another lesson learned from the Montreal Protocol is that international support for building capacities needs to precede the support to specific actions. The former enables recipient countries to fully participate in the phases of implementation and monitoring, and this has proved to be a necessary condition for the effective use of the transferred resources [Joffe *et al.*, 2013].

Finally, financial transfers can play a role not only as a "carrot" but also as a "stick". It has been argued that a credible system of sanctions in case of non-compliance constitutes a crucial element for an environmental agreement to be successful [Barrett, 2008; de Zeeuw, 2015; Edenhofer *et al.*, 2014]. The temporary suspension of transfers in case of non-compliance can be a credible sanction. Moreover, some authors point out that sanctions do not negatively affect but enhance the fairness perception about the agreement if the other two elements (burden-sharing rules and international transfers) are fairly designed [Barrett, 2010].

4. International Negotiations and the Current and Future Global Climate Finance Architecture

In the previous two sections, we discussed the magnitude of the climate finance challenges ahead and the role of financial transfers in international climate negotiations. Here we will describe the complexity and evolving nature of the global climate finance architecture that has emerged during the long process of international climate negotiations and that has led the international community to a climate finance standoff.

Since the negotiation of the UNFCCC in 1992, the number of climate change funds has increased rapidly. Today, there are around 100 international public funds, including the Global Environment Facility (GEF), the Adaptation Fund, the Climate Investment Funds (CIFs), and most recently the Green Climate Fund (GCF), as well as a great number of private funds. Some reports [EC, 2009] argue that a decentralized, bottom-up system based on country-driven proposals allows contributors a higher degree of discretion over the efficient use of their contributions and is thus likely to encourage larger contributions than could be expected in the case of only one major centrally managed multilateral fund. However, other reports argue that the existing architecture for international climate finance is overly complex and fragmented, and suggest that a more holistic approach could be beneficial [CFSG, 2015]. The issues of fragmentation and over-complexity are discussed in more detail below.

At the climate conferences in Copenhagen (2009) and Cancun (2010), developed countries pledged jointly to provide nearly US$30 billion in FSF to developing countries in 2010–2012 to support immediate action on the ground, and to mobilize US$100 billion[5] a year by 2020 from a variety of sources. At the recent conference in Paris (December 2015), this amount was confirmed as the floor of the collective goal of funding that developed countries assume to transfer to developing countries to support them in the transition to low carbon resilient economies [COP 21, 2015]. These commitments were seen at the time as a breakthrough in the international climate finance regime [The Economist, 2010], but the vague nature of the target and the slow progress in meeting this goal have led to equity concerns being particularly prominent in discussions of climate finance and may pose a challenge to reaching successful international cooperation in climate policy [Boyd *et al.*, 2015].

The scale of the US$100 billion commitment could be influenced to some extent by the available scientific estimates of the costs of adaptation and mitigation in developing countries described in Section 2 of this chapter. However, it is almost certain to be too low a target if it is to comprise both public and private sources of climate finance and too little if it is to be used for both adaptation and mitigation. Thus, this leads us to think that the overall target for developed countries was determined essentially on a political basis and that the little clarity on rules for what counts as climate finance and country-specific commitments and entitlements helped reach an agreement but also served to perpetuate the unbalances (public/private finance, adaptation/mitigation finance, climate/development finance) that prevent the international community to deliver effective investment at scale in both mitigation and adaptation.

Transparency, adequacy, additionality, geographic and thematic balance, equal access and effectiveness are important challenges that

[5]It is expected that a significant amount of future international climate funding will be channeled through the GCF, which became operational in May 2014.

a post-Paris agreement should address with regard to international climate finance. In what follows we will discuss them separately.

The effective measurement, reporting and verification of climate finance are key factors for building trust between Parties to the Convention, and also for external actors. The absence of a comprehensive system for tracking climate finance, and the lack of clear definitions for climate finance, mitigation and adaptation projects under the UNFCCC, together with the low transparency, impede an accurate tracking of climate finance flows. Data have to be compiled from different sources, with heterogeneous quality, timeliness, assumptions, coverage and methodologies. Besides, there are information gaps, especially regarding private investments in adaptation, forestry and energy efficiency. Most data correspond to commitments rather than to disbursements (the progress in disbursements has been slow), reflect the total cost of the investment rather than the portion attributed to climate actions, and include different types of finance, although most are concessional loans or grants [Buchner *et al.*, 2015; Fridahl and Linnér, 2015; Mansell, 2015; UNFCCC, 2014]. Efforts are in place to improve climate finance tracking. For example, an OECD-led network of research organizations, international finance institutions, and governments is currently involved in the development of methodologies for estimating private finance mobilized by developed countries' public interventions towards low-carbon and climate-resilient activities in developing countries [Jachnik *et al.*, 2015]. Scaling up and shifting public and private sector investments into low-carbon, climate resilient infrastructure in developing countries will require strengthened international efforts to track North-to-South climate finance.

According to the most recent tracking initiatives, in 2010 approximately 53–60% of North-to-South climate finance came from private sources [Gurría, 2014], and the volume of public funding was between US$35–49 billion per year for the period 2011–2012, and between US$37.9–43.5 billion for the period 2013–2014 [Edenhofer *et al.*, 2014; OECD, 2015a]. The stagnation of overall amounts of finance in 2011–2012 levels indicates that the trajectory of climate finance is not aligned with needs, something that would question the ability of developed countries to mobilize resources at the adequate scale if transfers follow the same pattern in the future [Boyle *et al.*, 2014].

Several authors have also questioned whether the transferred resources have been actually new and additional, as specified in the commitment by developed countries. According to UNFCCC guidelines, countries should indicate why they consider their contributions meet the criteria of being new and additional. Each country, however, interprets additionality in a different manner and the fact that non-climate-related bilateral aid funding has declined at the same time that climate-related aid has increased is a powerful argument for those who claim that climate finance has crowded out official development assistance [Edenhofer *et al.*, 2014; Kaul *et al.*, 2014; UNFCCC, 2014].

The failure to mobilize additional funding by developed countries becomes a concern for the neediest countries after observing the pattern of climate finance geographical and thematic distribution. So far, public finance has predominantly focused on mitigation and funding disbursements for mitigation have concentrated in a small number of emerging economies with high mitigation potential (China, India, South Africa and Brazil). Private finance has also focused almost entirely on mitigation projects in emerging countries. Adaptation finance, which comes almost exclusively from public sources, has increased in the past few years but still represents a small share of total climate finance [Buchner *et al.*, 2015]. Thus, low-income countries in the Middle East and Africa, where adaptation needs are concentrated and mitigation potential is low, have problems to access climate finance [Boyle *et al.*, 2014; UNFCCC, 2014]. Many developing countries also lack the technical capacity to deal with the bureaucratic burdens associated with the proliferation of diverse funding channels with different governance procedures and eligibility conditions [UNFCCC, 2014]. These countries fear being left behind as traditional aid funding is diverted from development needs towards mitigation finance in emerging countries, as private finance becomes more important, and as the complexity of climate finance landscape continues increasing [Boyle *et al.*, 2014].

The fact that adaptation projects are primarily focused on addressing other development objectives might explain the demand of developing countries for an equal treatment for mitigation and adaptation objectives [OECD, 2015b]. As stated by Fridahl and Linner [2015], these countries are very interested in the GCF to become the main channel of long-term finance, as it will seek a balanced allocation of resources between mitigation and adaptation. These authors also argue that developed countries may see mitigation as a more urgent issue than adaptation, and therefore that the current bias towards mitigation could be interpreted as a consequence of donors' preferences.

Improving accessibility might also be important to the donors. According to Glemarec [2011], if only emerging economies had access to mitigation finance, developed and emerging countries would have to achieve very high emissions reductions in 2050 (up to 95% and 75% of 1990 levels, respectively). Thus, making finance accessible to the least developed countries may help reduce the mitigation burden of developed and emerging countries.

Finally, effectiveness of climate finance has been discussed at length but unfortunately to date estimates of the actual effectiveness are very few. The Intergovernmental Panel on Climate Change (IPCC) recognized that evidence of the effectiveness of climate finance is lacking and points to the complexity of the climate finance landscape as a possible reason. According to the IPCC, effectiveness of climate finance would increase if developing countries had capacity to cope with the diversity of channels, and to manage the available sources of finance in favor of their national development strategies and needs [Edenhofer *et al.*, 2014]. The relationship between effectiveness and country ownership was already recognized in the 2005 Paris Declaration on Aid Effectiveness and the Accra High-Level Forum of Aid Effectiveness [Gomez-Echeverri, 2013].

5. Conclusions

Climate finance will be an important element for achieving the target set in the agreement adopted on 12 December 2015 in Paris of limiting the global average temperature increase to well below 2°C above pre-industrial levels, and for achieving a low carbon, climate-resilient sustainable development. In Copenhagen in 2009, the North agreed that US$100 billion per year should flow to the South in climate finance by 2020. The Paris Agreement makes it clear that developed countries will continue to mobilize US$100 billion per year from 2020 to 2025, and that by 2025 countries will set a new collective goal for climate finance above US$100 billion per year.

Many experts claim that the US$100 billion figure is just a symbolic number "picked out of the air" by governments at Copenhagen to get a last-minute deal [The Guardian, 2015]. Estimates suggest that the annual costs for climate adaptation and loss and damage alone in developing countries are already well in excess of US$100 billion. Thus, it is highly unlikely that even with US$100 billion in North-to-South transfers developed countries will meet the goal of covering the agreed full incremental costs of mitigation measures and the essential adaptation needs of poor and marginalized people in developing countries.

The agreed figure could be seen, however, as an important political goal that will signal to the private sector the developed countries' commitment to scaling up climate finance. It will also constitute a "totemic" issue for the least developed countries that have little or no responsibility for climate change but are the ones most affected by and most vulnerable to its impacts. Both reasons put extra urgency in closing the funding gap by 2020 as a way to avoid the "trust deficit" that would weaken the implementation of the Paris Agreement and reduce the capacity to leverage private sector investment in climate-relevant projects.

The challenge of rebuilding trust implies not only legally binding individual countries' commitments but also clarity around what sources of finance can be included to meet the 2020 commitment.[6] So far, developed countries have not accepted a fully-fledged top-down approach for determining contributions to climate finance. In the case of FSF, the unilaterally determined contributions of countries were sufficient for reaching the collective goal of US$30 billion. Donors could reach the collective goal on that occasion because some contributions (especially the Japanese one) were large and the scale of the goal was assumable, as well as because a portion of the funds was not new and additional, but diverted from aid funding for other development purposes. Given that long-term finance goal is 10 times larger, unilaterally determined contributions might fall short to reach it.[7]

Some interesting lessons can be learned from the negotiations on mitigation efforts. Up to 2009, discussions about such efforts at the UNFCCC took a top-down approach and tried to apply various sets of equity criteria to distribute the "burden" among individual countries. One of the main factors that explains the lack of success of such an approach (apart from the reluctance of parties to accept stringent mitigation commitments and control of their achievement by an international institution and the clear incentive for individuals as well as states to free-ride on other's mitigation effort) is that there is no single agreed distributional justice principle that can be used to construct a proposal with a sufficiently high degree of legitimacy.[8]

[6]Transparent climate finance flows may also help all stakeholders to identify opportunities to further mobilize or use finance more effectively.

[7]Pickering *et al.* [2015] suggest that if countries could agree in one particular criterion (e.g., GDP) to determine the individual contribution of each country, then the collective goal could be achieved. Otherwise, the gap could be around 15% if countries could unilaterally decide between responsibility (historical emissions) and capacity (GDP) criteria, or much larger if decisions were based in other considerations (e.g. previous contributions to climate finance). This gap is even greater when the set of contributing countries extends to emerging countries.

[8]In recent years, a number of proposals have emerged based on the principles of equality, responsibility, capacity and need but even in these cases there is very little agreement about how exactly these principles should be applied and therefore they are subject to self-serving interpretation by each country.

However, recent studies such as Averchencova *et al.* [2014] show that what really determines the distribution of mitigation efforts across countries is the ambition of the global target and the growth in economic size of each region.

In any case, to overcome the standoff in climate negotiations the post-2009 mitigation policy at the UN has shifted focus to a hybrid approach, a sort of "pledge and review" system that takes as its point of departure the so-called Intended Nationally Determined Contributions (INDCs) where countries will have to set out mitigation commitments and clear plans on how to meet them before the Paris meeting.

This new approach has important implications with regard to climate finance. The INDCs (together with the National Adaptation Plans (NAPs)) will be the basis to guide the design of national infrastructure development plans, and will give the international community a true sense of what is needed and when. While it is challenging to determine precise needs, the exact amount and composition of finance provided and mobilized will depend on national circumstances in recipient countries. Thus, a country-by-country assessment (revised over time) will be a useful tool for a bottom-up identification of finance needs that will then help build financing schedules for both developed and developing countries. Matching pledges to specific mitigation and adaptation plans may provide the transparency and predictability required to build cooperation and collaboration between contributors and recipients and drive up an upward spiral of greater ambition. The suggested approach would also bring other co-benefits for international climate negotiations: (1) it would allow for the negotiation of financing schedules where top priority would be given to institutional capacity-building in recipient countries; (2) it would make climate finance an integrated part of the effort-sharing equation to be negotiated, linking for every region domestic emission reduction plans and support required for/offered to low carbon development in developing regions; and (3) it could also spur South-to-South solidarity in climate finance and relieve some of the reluctance of developed regions that ask for commensurate actions of emerging economies before committing unilaterally to mitigation and financial efforts.

In any case, many challenges will still remain. First, the donors' incentive for an imbalanced allocation of climate finance between mitigation and adaptation seems to suggest that it may be necessary to include in negotiations a separate public finance target for adaptation. Second, it will also be necessary to agree on a common and transparent definition of additionality of climate finance so that we avoid double counting of development aid as climate finance and the associated lack of trust in international negotiations. Third, the growing distinction between developing countries and emerging economies makes the use of a simple North-South classification of countries difficult to maintain. All these will require working with some baseline concessional financing mobilized by donor countries.

Acknowledgments: The participation of Maria Victoria Román in this chapter has been funded by CICEP (Strategic Challenges in International Climate and Energy Policy), one of three centers for social science based research on environment-friendly energy established by the Research Council of Norway in 2011 (http://www.cicep.uio.no/).

References

Abadie, L.M., I. Galarraga and D. Rübbelke. 2013. An Analysis of the Causes of the Mitigation Bias in International Climate Finance. *Mitigation and Adaptation Strategies for Global Change* 1–13.

Averchenkova, A., N. Stern and D. Zenghelis. 2014. *Taming the Beasts of "Burden-Sharing": An Analysis of Equitable Mitigation Actions and Approaches to 2030 Mitigation Pledges.* Policy Paper. Centre for Climate Change Economics and Policy and Grantham Research Institute on Climate Change and the Environment. At: http://www.lse.ac.uk/GranthamInstitute/publication/taming-the-beasts-of-burden- sharing- an- analysis-of-equitable-mitigation-actions-and- approaches-to-2030- mitigation-pledges/.

Barrett, S. 2008. Climate Change Negotiations Reconsidered. *Progress on Agenda for Global Action Strategy* 93.

Barrett, S. 2010. *Negotiating the Next Climate Change Treaty.* Policy Exchange. London, UK.

Bodansky, D. and E. Diringer. 2014. Alternative Models for the 2015 Climate Change Agreement. *FNI Climate Policy Perspectives* 13.

Bowen, A., E. Campiglio, and S. Herreras. 2015. *The "Optimal and Equitable" Climate Finance Gap.* Centre for Climate Change Economics and Policy. Working Paper No. 209. At: http://www.lse.ac.uk/GranthamInstitute/wp-content/uploads/2015/03/Working-Paper-184-Bowen-et-al.pdf.

Boyd, R., F. Green and N. Stern. 2015. *The Road to Paris and Beyond.* Policy Paper. August. Centre for Climate Change Economics and Policy Grantham Research Institute on Climate Change and the Environment. At: http://www.lse.ac.uk/GranthamInstitute/wp-content/uploads/2015/08/The-road-to-Paris-and-beyond.pdf.

Boyle, J., D. Runnalls and S. Sawyer. 2014. *Institutional Arrangements for Climate Finance.* CIGI Papers No. 26. March 2014. Centre for International Governance Innovation (CIGI). Waterloo, Canada. At: https://www.cigionline.org/sites/default/files/no26_0.pdf.

Buchner, B., M. Stadelmann, J. Wilkinson, F. Mazza, A. Rosenberg and D. Abramskiehn. 2014. *The Global Landscape of Climate Finance 2014.* Climate Policy Initiative (CPI).

Buchner, B., C. Trabacchi, F. Mazza, D. Abramskiehn and D. Wang. 2015. *Global Landscape of Climate Finance 2015.* Climate Policy Initiative (CPI). At: http://climatepolicyinitiative.org/wp-content/uploads/2015/11/Global-Landscape-of-Climate-Finance-2015.pdf

G20/CFSG. 2015. *Report to the Finance Ministers.* G20 Climate Finance Study Group. At: http://nbsapforum.net/uploads/1614.pdf.

Ciplet, D., J. T. Roberts and M. Kahn. 2013. The Politics of International Climate Adaptation Funding: Justice and Divisions in the Greenhouse. *Global Environmental Politics* 13, 49–68.

Conference of Parties 21 (COP 21). 2015. *Adoption of the Paris Agreement.* Proposal by the President. Draft decision -/CP.21.

Dellink, R. B., M. Den Elzen, H. Aiking, E. Bergsma, F. Berkhout, T. Dekker and J. Gupta. 2009. Sharing the Burden of Financing Adaptation to Climate Change. *Global Environmental Change* 19, 411–421.

de Zeeuw, A. 2015. International Environmental Agreements. *Annual Review of Resource Economics* 7, 1.1–1.18.

EC. 2009. *Stepping up International Climate Finance: A European Blueprint for the Copenhagen Deal.* Communication from The Commission to the European Parliament, The Council, The European Economic and Social Committee and The Committee of the Regions.

Edenhofer, O., R. Pichs-Madruga, Y. Sokona, E. Farahani, S. Kadner, K. Seyboth, A. Adler, I. Baum, S. Brunner and P. Eickemeier. 2014. *Climate Change 2014: Mitigation of Climate Change.* Contribution of Working Group III to the Fifth Assessment Representative Intergovernmental Panel Climate Change, 511–597.

Fridahl, M. and B.-O Linnér. 2015. Perspectives on the Green Climate Fund: Possible Compromises on Capitalization and Balanced Allocation. *Climate and Development* 1–5.

Gichira, P. S., J. F. Agwata and K. D. Muigua. 2014. Climate Finance: Fears and Hopes for Developing Countries. *Journal of Law, Policy and Globalization* 22, 1–7.

Glemarec, Y. 2011. *Catalyzing Climate Finance: A Guidebook on Policy and Financing Options to Support Green, Low-Emission and Climate-Resilient Development.* United Nations Development Programme. New York, NY.

Gomez-Echeverri, L. 2013. The Changing Geopolitics of Climate Change Finance. *Climate Policy* 13, 632–648.

Grasso, M. 2010. An Ethical Approach to Climate Adaptation Finance. *Global Environmental Change* 20, 74–81.

Gurría, A. 2014. *Improving Transparency and Accountability through Improved Tracking of Climate Finance Flows* [Internet]; [updated 2015 December 30]. Paris: OECD. At: http://www.oecd.org/about/secretary-general/improving-transparency-and-accountability-through-improved-tracking-of-climate-finance-flows.htm.

Höhne, N., M. Den Elzen and D. Escalante. 2014. Regional GHG Reduction Targets Based on Effort Sharing: A Comparison of Studies. *Climate Policy* 14, 122–147.

Jachnik, R., R. Caruso and A. Srivastava. 2015. *Estimating Mobilised Private Climate Finance: Methodological Approaches, Options and Trade-offs.* OECD Environment Working Papers No. 83. OECD Publishing. At: http://dx.doi.org/10.1787/5js4x001rqf8-en.

Joffe, P., D. Waskow, K. DeAngelis, W. Bevins and Y. Dagnet. 2013. Equity Lessons from Multilateral Regimes for the New Climate Agreement. World Resources Institute.

Kaul, I., D. Blondin and N. Nahtigal. 2014. *Country Allocations versus Issue Allocations: The Case of Climate Finance.* Discussion Draft (2014) now published in: M. Boussichas and P. Guillaumont, eds. (2015). *Financing Sustainable Development: Addressing Vulnerabilities.* FERDI: Clermont-Ferrand, pp. 353–370. At: http://www.ingekaul.net/wp-content/uploads/2015/06/CountryAllocationsVersusIssueAllocations_PDF_2.pdf.

Mansell, A. 2015. *Addressing Finance in a 2015 Climate Agreement.* Center for Climate and Energy Solutions (C2ES).

Organisation for Economic Co-operation and Development (OECD). 2015a. *Climate Finance in 2013–14 and the USD 100 Billion Goal.* A report by the OECD in collaboration with Climate Policy Initiative (CPI).

Organisation for Economic Co-operation and Development (OECD/DAC). 2015b. *Climate-Related Development Finance in 2013: Improving the Statistical Picture.* The OECD Development Assistance Committee. At: http://www.oecd.org/dac/environment-development/Climate-related%20development%20finance%20FINAL.pdf.

Organisation for Economic Co-operation and Development (OECD)/International Energy Agency (IEA). 2015. *Energy and Climate Change.* World Energy Outlook Special Report.

Pickering, J., J. Skovgaard, S. Kim, H. Reich, J. T. Roberts, D. Rosatti and M. Stadelmann. 2015. Acting on Climate Finance Pledges: Inter-Agency Dynamics in Contributor States. *World Development* 68, 149–162.

Pittel, K. and D. Rübbelke. 2013. International Climate Finance and Its Influence on Fairness and Policy. *The World Economy* 36, 419–436.

Ringius, L., P. Frederiksen and K. Birr-Pedersen. 2002. *Burden Sharing in the Context of Global Climate Change: A North-South Perspective.* NERI Technical Report No. 424, 90 pp. National Environmental Research Institute. Denmark. At: http://technical-reports.dmu.dk.

Schmidtz, D. 2006. *The Elements of Justice.* Cambridge, UK: Cambridge University Press.

Stadelmann, M., A. Michaelowa and J. T. Roberts. 2013. Difficulties in Accounting for Private Finance in International Climate Policy. *Climate Policy* 13, 718–737.

The Economist. 2010. *Climate Change: A Surprising Success.* The Economist [Internet]; [updated 2015 December 30]. At: http://www.economist.com/blogs/newsbook/2010/12/climate_change.

The Guardian, 2015. *Paris Climate Talks Should Not Put Figure on Finance, Says World Bank Vice-President.* The Guardian [Internet]; [updated 2015 December 30]. At: http://www.theguardian.com/environment/2015/oct/07/paris-talks-un-cop-climate-finance-world-bank-vice-president.

Tian, H. and J. Whalley. 2010. Trade Sanctions, Financial Transfers and BRIC Participation in Global Climate Change Negotiations. *Journal of Policy Modeling* 32, 47–63, doi: 10.1016/j.jpolmod.2009.09.004.

United Nations (UN)/Department of Economic and Social Affairs (DESA). 2009. Promoting Development, Saving the Planet. DESA. ST/ESA/319 No. E/2009/SO/Rev.1.

United Nations Environment Programme (UNEP). 2014. *The Adaptation Gap Report 2014.* UNEP. Nairobi, Kenya.

United Nations Framework Convention on Climate Change (UNFCCC). 2014. *2014 Biennial Assessment and Overview of Climate Finance Flows Report.* UNFCCC Standing Committee on Finance. At: http://unfccc.int/files/cooperation_and_support/financial_mechanism/standing_committee/application/pdf/2014_biennial_assessment_and_overview_of_climate_finance_flows_report_web.pdf.

van Melle, T., N. Höhne and M. Ward. 2011. *International Climate Financing: From Cancún to a 2° C Stabilisation Pathway.* Ecofys, by order of KfW Development Bank. At: http://www.ecofys.com/files/files/climate_financing_after_cancun_20110204.pdf.

Watkiss, P., F. Baarsch, N. Canales-Trujillo, A. Caravani, S. Nakhooda, S. Ockenden, M. Stadelmann and C. Trabacchi. 2014. *The Adaptation Funding Gap.* In UNEP Adaptation Gap Report, UNEP.

World Bank. 2010. *Economics of Adaptation to Climate Change.* Synthesis Report. World Bank.

Chapter 3

Adaptation Finance: Setting the Ground for Post-Paris Action

Chiara Trabacchi

Climate Policy Initiative,
20 St Dunstans Hill, London, EC3R 8HL, United Kingdom
chiara.trabacchi@cpiclimatefinance.org

Barbara K. Buchner

Climate Policy Initiative,
235 Montgomery Street, 13th Floor, San Francisco, CA
barbara.buchner@cpiclimatefinance.org

Abstract

Adaptation finance has been recognized as a pivotal element of the international response to climate change in light of the climate vulnerability of natural and human systems to climate-related risks in an increasingly warm world. At present, however, relevant data on adaptation financing are not systematically and consistently collected across investors, and there is not yet an internally agreed methodology for comprehensive and consistent measurement. The resulting information gap on adaptation finance flows is an impediment to accountability and the provision of quantitative-based evidence supporting adaptation policy and investment decision-making.

This study sought to provide an updated picture on how much, how, where, and from whom finance is flowing toward climate-resilient actions with the overarching aim of providing a pre-Paris Agreement baseline against which measuring progress and on which building to further improve tracking systems. Drowning from a variety of primary and secondary sources, the study finds that in 2013/2014 public institutions invested US$25 billion, on average, for adaptation actions. The study highlights that this estimate remains uncertain and partial due to lack of harmonized accounting methods and data gaps, particularly on private investment. The study points to improvement in current tracking systems and highlights opportunities for scaling up adaptation finance given that more efforts would be needed to enable societies and economies to adapt to the impacts of climate change.

1. Introduction

Adaptation finance — financial support for adaptation activities — has been recognized as a pivotal element of the international community's response to climate change since the outset of the climate change negotiations. The United Nations Framework Convention on Climate Change (UNFCCC), in fact, established developed countries' obligations to "*[...] assist the developing country Parties that are particularly vulnerable to the adverse effects of climate change in meeting costs of adaptation to those adverse effects*", by providing "*[...] new and additional resources to meet the agreed full incremental costs*" associated with "*preparing for adaptation to the impacts of climate change*" [UNFCCC, 1992, Article 4.4].

In 2010, the Cancun Conference of the Parties (COP) set another significant milestone in the climate finance discourse with relevance for adaptation. In the so-called Cancun Agreements, developed countries formally committed to the goal of mobilizing jointly USD 100 billion a year by 2020 to address the needs of developing countries. The Agreement also requests developed countries to provide developing countries with scaled-up, predictable, new and additional finance to implement urgent, short-, medium- and long-term adaptation actions, plans, programs and projects.

The Paris Agreement signed at the 21st COP in December 2015 places enhanced emphasis on adaptation, seeking to strengthen adaptation efforts under the Convention. Albeit it did not fix a dedicated quantitative financial target as many requested, the Agreement urges developed countries to "*significantly increase adaptation finance from current levels*". In the context of the USD 100 billion annual goal, it extended the timeframe from 2020 to 2025; a higher funding target will be set after 2025. The Paris Accord also recognized that adaptation is a global challenge faced by all with local, subnational, national, regional and international dimensions. In a world on a 4°C warming trajectory [IPCC, 2013; WB, 2012], a targeted and efficient use of limited public resources to mobilize private capital is critical to tackle such challenge.

Against this background, and building on Buchner *et al.* [2015] and Trabacchi *et al.* [2015], this study aims to:

i. Provide an updated picture on how much, how, where, and from whom, finance is flowing toward climate-resilient actions globally, thereby developing a pre-Paris Agreement baseline on international financial flows invested in pursuit of climate resilience objectives in developing countries.
ii. Highlight opportunities for scaling up private investment in climate resilience.

A proper measurement, tracking, and reporting system of relevant financial flows across a variety of sources is a critical building block to ensure finance is used efficiently and targeted where it is needed the most. It is also critical to understand the links between public and private finance thereby comprehending how public resources could be best used to catalyze greater private investment in activities that increase countries', businesses', and communities' resilience.

1.1. *The definitional issue*

The Convention does not operationally define the terminology related to central concepts regarding the provision of financial resources. The language used leaves ample room for interpretations [Machado-Filho, 2011], and its practical meaning has been the source of substantial debate and controversy [see Corfee-Morlot *et al.*, 2009; Romani and Stern, 2011]. Parties' divergent political views on how much should be made available, to whom, and from which source(s), have made it difficult to achieve an agreement on a definition of climate finance and to establish the basis for financial accounting rules or a measurement system for tracking relevant flows [Buchner *et al.*, 2011a, b].

Terminological and methodological challenges have hindered practitioners' ability to measure progress against the commitments made under the aegis of the UNFCCC (e.g., Article 4, UNFCCC, 1992; UNFCCC, 2010, 1/CP.16), as well as in terms of resource management to deliver on national policies. This is particularly true for adaptation finance.

First, a commonly agreed, concrete and discriminating definition of adaptation is missing under the UNFCCC [Persson *et al.*, 2009]. The Convention, in fact, has not specified which activities and which type(s) of financial flows should account toward the USD 100 billion goal, and how to systematically track all relevant climate flows. It is worth noting that the findings of country-level studies, such as Ampri *et al.* [2014], Caravani *et al.* [2014], Terpstra *et al.* [2013], and United Nations Development Programme (UNDP) [2013], also raise questions regarding who should determine what constitutes adaptation-relevant [Caravani *et al.*, 2014]. Donors and development partners' perspectives on what constitutes adaptation are, in fact, often different from those of recipient countries [Caravani *et al.*, 2014].

Second, adaptation is intricately intertwined with development, making difficult to distinguish it from other forms of development finance and to identify the "incremental costs" referred to in the Convention [Brown and Kaur, 2009]. Many adaptation activities, in fact, overlap with — and should be mainstreamed into — traditional development activities [Buchner *et al.*, 2011b], and many conventional development interventions (such as social protection or disaster risk reduction programs) can influence the ability of communities to adapt [Levine *et al.*, 2011].

Third, adaptation takes place in response to multiple stimuli. It is therefore conceptually and practically problematic to identify whether adaptation takes place in response to development pressures or to changes in climatic conditions [see Berrang-Ford *et al.*, 2011; Levine *et al.*, 2011]. Ultimately, the characteristics of socioeconomic systems influence the degree of possible climate change risks, as determinants of exposure and sensitivity to climate change effects, and systems' ability to adapt.

Adaptation, in fact, is highly contextualized within a system [Smit and Pilifosova, 2002] and, therefore, dependent upon context-specific factors including climatic, institutional, and socio-economic ones. As such, an activity supporting adaptation objectives in one context may result in "maladaptation" in another [Tirpak *et al.*, 2010].[1]

Fourth, adaptation is a dynamic phenomenon, evolving over time in response to a changing climate [Hallegatte, 2009], and context-specific conditions. Arnell [2011], in fact, points to an iterative climate risk management approach. For this reason, adaptation measures not only refer to activities tackling underlying drivers of vulnerability, by empowering socio-economic systems to respond to current climate variability and risks, which may or may not be directly related to climate change; but also activities enabling to address future climate conditions by incorporating future climate scenario into investment decisions.

In addition to technical challenges, the political economy of adaptation can also influence the process of tracking adaptation finance, which is inherently political [Jones, 2012; Michaelowa and Michaelowa, 2011].

These characteristics of adaptation represent challenges to the identification of what types of activities should be tracked and how should be accounted against international commitments. They represent challenges to both donor and recipient countries, which are requested to report on the support provided and on the

[1]For instance, switching crop species to offset the negative effects of climate change on yields is an adaptation strategy that cannot be blindly replicated across contexts because it is highly context-specific and dependent on e.g. quality levels of soil, local weather etc. Another example is offered by Field *et al.* [2014] in the context of adaptation strategies in small island countries. Field *et al.* [2014] flag that albeit lessons learned from adaptation experiences in one island may offer guidance for measures in others, cautions must be exercised to ensure that such measures are appropriate to local biophysical, social, economic, political and cultural circumstances to avoid inappropriate or unintended results.

support received [see UNFCCC, 2010; 1/CP.16 and UNFCCC, 2011, 2/CP.17].

For recipient developing countries, the challenges can be even higher. Donor countries and institutions, in fact, use a variety of definitions and criteria to identify climate finance, as well as different classifications for type of activities [Tirpack *et al.*, 2012]. Decision-makers in these countries need a complete and clear picture as well as accurate data to coordinate and manage funds from across multiple organizations, and fundraise for national climate change-related activities [Tirpack *et al.*, 2012].

The lack of clear definitional boundaries under the UNFCCC, and of consensus on methodologies, has so far hindered understanding. The government bodies in charge of reporting on adaptation spending find the tracking process confusing and cumbersome due to the ambiguity in classification criteria [Bhattarai *et al.*, 2011; Jones *et al.*, 2012], thereby not useful to plan expenditure.

The Paris Agreement aims to address the definitional and tracking issues by requesting to the Subsidiary Body for Scientific and Technological Advice to develop modalities for the accounting of financial resources provided. This is an important step to ensure efficient and effective management of public resources.

2. Methodological Approach

To collect and aggregate empirical data on annual financial flows supporting climate resilience interventions, the study adopts and builds on the accounting framework developed in the context of the Global Landscape of Climate Finance series [see Buchner *et al.*, 2011a, 2012, 2013 and 2014].

In particular, in the absence of an internationally agreed definition of what qualifies as adaptation finance, the study relies on the tracking and accounting methodologies used by: i) the members of the OECD's Development Assistance Committee (DAC), the so called Rio Markers; ii) the group of Multilateral Development Banks (MDB) jointly reporting on climate finance; iii) the members of the International Development Finance Club (IDFC);[2] and Climate Funds.

The scope of the analysis is limited to:

- Capital flows directed towards activities aimed at reducing the vulnerability of human or natural systems to the impacts of climate change and climate-related risks.[3] These flows include support for capacity building measures as well as for the development and implementation of policies.[4]
- Public flows, namely commitments from Development Finance Institutions (DFIs), governments and their agencies and Climate Funds. Private flows are not covered due to data availability issues, unresolved methodological difficulties and the lack of adequate reporting systems [see Buchner *et al.*, 2012, 2013 and Clapp *et al.*, 2012].

The study captures:

- Annual financial commitments in the latest available year provided through grants, low-cost (including concessional) and market term loans, project-level equity, and balance sheet financing (i.e., a direct debt or equity investment by a company or finance institution).[5]

[2]See OECD [2011]; AfDB *et al.* [2012a, 2012b, 2013, 2014 and 2015]; AfDB [2015]; WBG [2015a]; Höhne *et al.*, [2012]; and IDFC [2013 and 2014].

[3]It is worth noting that data collected from the group of Multilateral Development Banks (MDBs) jointly reporting on climate finance reflect their process-based approach to adaptation finance tracking, which is context- and location-specific, conservative and granular. In July 2015, these MDBs and the members of the IDFC established "Common Principles for Tracking Adaptation Finance" and the OECD DAC has been refining the Rio Marker definition to reflect the MDBs' principles.

[4]Our working definition of climate finance is aligned with the recommended operational definition of the UNFCCC Standing Committee of Finance [see UNFCCC SCF, 2014] which states: "*Climate finance aims at reducing emissions, and enhancing sinks of greenhouse gases and aims at reducing vulnerability of, and maintaining and increasing the resilience of, human and ecological systems to negative climate change impacts.*"

[5]The share of climate finance allocated to different categories of financial instruments may not fully reflect reality, as our categorization is based on the quality of the data sources we can access.

- Total primary financial transactions and investment costs or, where tracked, components of activities that directly contribute to adaptation, plus public framework expenditures (e.g., development of national climate strategies).

While we acknowledge the importance of risk management instruments, we exclude these from the total climate finance figure to avoid double counting between, for example, the face value of full loan guarantees and loans.

The public versus private nature of flows is determined by the type of actor undertaking a given transaction. In alignment with OECD (2013), finance qualifies as public if carried out by central, state or local governments and their agencies at their own risk and responsibility.

In terms of recipients, the analysis maps the initial recipients of flows, namely the first entities receiving money from the source or intermediary of climate finance. To this end, to determine the public and/or private nature of the recipient we used information provided by donors in the OECD-DAC Creditor Reporting System (CRS) database, reporting from DFIs, or details available in Climate Funds data sets.

2.1. *Data collection and aggregation approach*

This research captures 2013 and 2014 data, the latest available at the time of undertaking the quantitative analysis (May-October 2015). The data that subsequently become available do not influence the main findings of this research.

Empirical data and information were gathered from a wide range of primary and secondary sources. The main sources used are:

1. The OECD-DAC CRS Aid Activities database [OECD, 2015], to identify governments' adaptation-related aid commitments;
2. The Climate Finance Update website [ODI and HBF, 2015] and official documents to collect and verify data on Climate Funds' commitments;

3. A financial survey ad hoc developed to collect comparable primary data on DFIs' own resources committed for climate-relevant projects in 2014, as not otherwise publicly or readily available; DFIs' annual reports and a data set compiled by IDFC [2014] complemented the data collected via the survey;
4. Third-party expertise as a complement to the desk-based assessment approach to gain clarifications on tracking methodologies, or suggestions when data and/or information were lacking or difficult to interpret.[6]

3. Results: The Landscape of Adaptation Finance

The research finds that in 2013/2014 public institutions committed at least USD 23–26 billion for activities with adaptation objectives in developing countries, USD 25 billion on average. About 90% of this was invested in developing countries.

This figure, that represented around 17% of all public climate finance invested in 2014 according to Buchner *et al.* [2015] is a partial and uncertain estimate as it is affected by the different accounting approaches used for tracking finance, and tracking gaps for domestic budgets and private investment.

When compared to previous years' estimates from Buchner *et al.* [2011, 2012, 2013 and 2014], total adaptation finance in 2013/2014 is roughly the same as in 2012/2013. However, notwithstanding some year-to-year fluctuations observed in annual commitments, data show that public climate finance for adaptation has steadily increased since Buchner *et al.* [2011] first attempted to estimate international adaptation finance flows.

Furthermore, there are two noteworthy developments to flag:

- First, as shown in Trabacchi and Mazza [2015], the number of private sector-oriented adaptation projects in DFIs' portfolios has also increased over the past five years, and progress has been

[6]In addition to interactions with data providers, a Steering Committee composed of experts from DFIs and research organizations provided expert advice during the development of the Buchner *et al.* [2013] project.

made in designing approaches and business models to drive private investment in climate resilience.
- Second, the adoption of Common Principles for Climate Change Adaptation Finance Tracking from a number of bilateral, national, and multilateral DFIs as well as the fine-tuning of the OECD's Rio marker definition to reflect the criteria applied by MDBs represent a significant milestone towards harmonization.

Figure 1 illustrates adaptation finance flows along their lifecycle for the latest available year. The diagram shows that DFIs raised, managed and distributed the majority of public resources, contributing USD 21 billion in 2014 — about 84% of total climate finance flows.

DFIs often blend own financing with Governments and Climate Funds' concessional resources. In our analysis, we captured USD 1 billion from Climate Funds and a further USD 3 billion on average of direct public contributions from governments, ministries, and bilateral agencies for climate projects, mostly directed to low- and middle-income countries.

Public concessional or lower-than-market-rate finance, including loans with longer tenors and grace periods, plays a catalytic role by supporting the establishment of policy frameworks, strengthening technical capacity, lowering investment costs, and reducing investment risks for the first movers in a market.

Overall, most of the USD 25 billion captured in our Landscape 2015 (around 72% or USD 18 billion) was extended in the form of low-cost project debt and grants. Market-rate project debt represented 26% of the total or about USD 6.6 billion.

The bulk of adaptation investments targeted the climate change vulnerabilities of countries in East Asia and the Pacific region, Sub-Saharan Africa, and Latin America and Caribbean, 46% (USD 12 billion), 13% (USD 3 billion) and 12% (USD 3 billion) of total adaptation finance respectively.

Interventions in the water and wastewater management sector attracted most of adaptation flows in 2014 (USD 14 billion), followed by the agriculture and land use sector (USD 3 billion). It is worth

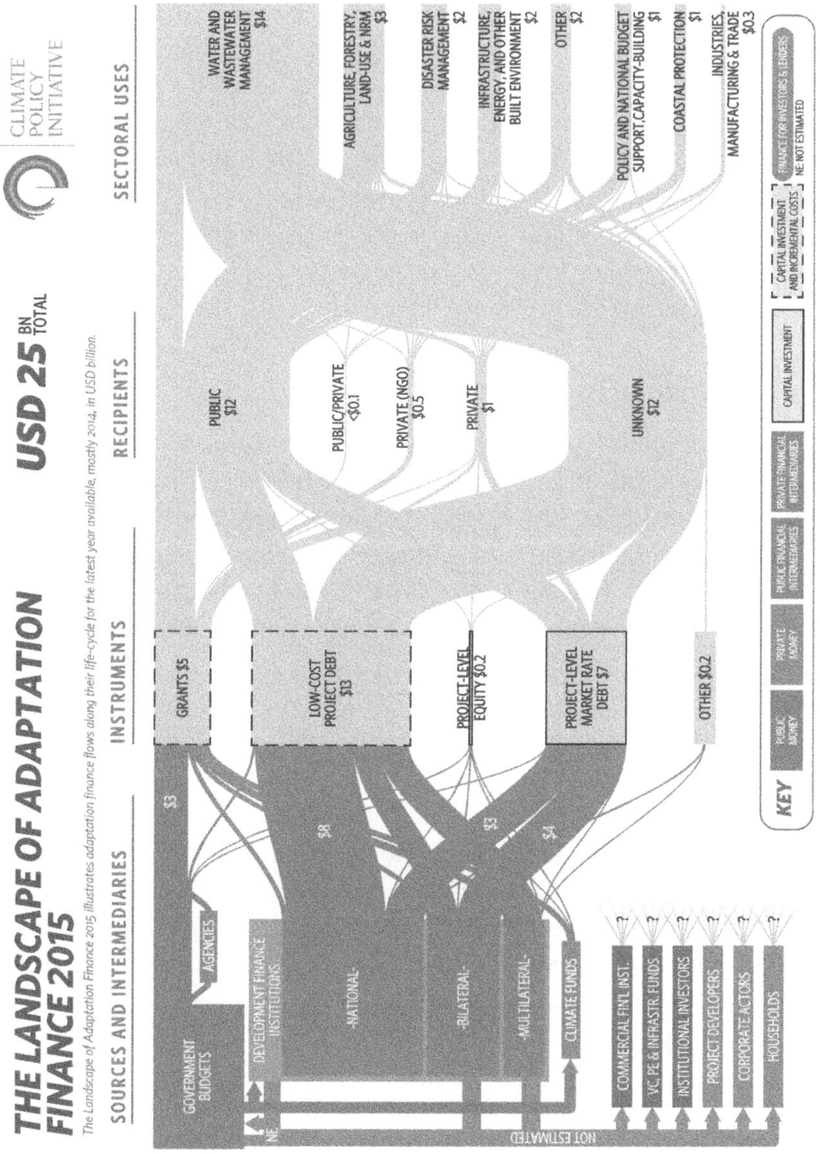

Figure 1. The Global Landscape of Adaptation Finance 2015.

noting that these are the key sectors prioritized in the adaptation components of the National Determined Contributions submitted by Parties ahead of the Paris climate summit [see AGR, 2015].

In terms of recipients, at least USD 1.4 billion or 21% of public climate finance for which we had details went to private entities (including non-governmental organisations (NGOs)) for adaptation projects. Details on recipients, however, are not yet consistently tracked or reported across data sources. Albeit our understanding of the links between public and private finance across the lifecycle of flows has progressed over the past years, it is still hindered by incomplete information.

3.1. *Taking stock of results*

A lot of work remains to be done to understand the actual size and nature of adaptation and how it is channeled on the ground.

Since Buchner *et al.* [2011], significant progress has been made in tracking relevant financial flows and getting a comprehensive picture of climate finance. The ongoing efforts across the climate finance tracking community such as the joint agreement on Common Principles for tracking adaptation finance by a group of multilateral, bilateral, national and sub-regional DFIs, have laid the foundation for further collaborative work to address outstanding comparability issues harmonize climate finance tracking methodologies.

Despite efforts, however, information about private investment in adaptation remains a key gap in the *Landscape*. Publicly-available project-level data for adaptation only provide an indication of a possible (not actual) private finance involvement[7] [Brown *et al.* 2015]. Private actors tend to consider climate resilience as part of their broader risk management processes and a source of competitive

[7] Evidence from the portfolio of MDBs highlights that USD 270 million of their financing made 26 private projects with a total value of USD 5.5 billion more climate-resilient [See VividEconomics, 2015]. A number of DFIs are developing methodologies to track private finance associated with their public climate finance interventions, covering both adaptation as well as mitigation finance.

advantage not to be disclosed. As a result, the private sector does not label and provide financial data on adaptation-related investments.

Beyond data on the volume of private investment, it is clear that what is missing the most are private investment to the level required to strengthening countries' resilience in a world on a 4°C warming trajectory and to achieve the (I)NDCs' goals. This is because private actors' action is hindered by a number of barriers that can affect their incentive and ability to invest in climate resilience.

Trabacchi *et al.* [2015] categorized such barriers as policy gaps, knowledge gaps, funding, and viability and risk coverage gaps. These refer to (i) non-existent or deficient regulatory frameworks that fail to put an adequate price on the risk of inaction, thereby lowering the rate of return of possible climate-resilient investment opportunities; (ii) the inability to recognize and evaluate the materiality of climate change risks and how to manage them; and (iii) inadequate access to finance or uncertainties about the returns of investments and risk aversion.

These barriers are particularly relevant for micro and small and medium-sized enterprises (SMEs) in middle-income and developing countries which typically have limited financial and technical capacity to carry out climate risk assessments and develop bankable projects.

Better information on private adaptation finance is an important gap to fill. Private sector engagement represents in fact a key opportunity for scaling up climate finance, an opportunity the public sector has to seize.

4. Concluding Remarks: The Way Forward

With at least USD 25 billion invested in climate-resilient actions globally in 2014 on average, public adaptation finance steadily increased over the past years. Nevertheless, current efforts need to expand to achieve the global goal on adaptation established in the Paris Agreement.

The private sector has a particular role to play in leveraging resources and expertise, and reaching solutions faster and more effectively. For the public sector, tackling those barriers inhibiting greater private engagement in climate resilience is a more urgent task than ever to support countries to efficiently transition towards a low-carbon and climate-resilient future.

The good news is that there are opportunities to be seized to this end for scaling up adaptation finance.

First, enhance tracking efforts to further improve the transparency, comprehensiveness and consistency in accounting approaches. This is important because a better understanding on how public climate finance is invested on the ground is a critical building block to ensure finance is used efficiently and targeted where it is needed the most. A clear understanding of where investments take place and what their underlying drivers are, can help to identify and overcome the remaining barriers to private investment.

Second, adjust regulatory frameworks to create stronger incentives for private investment. DFIs and other public agencies could help businesses and policy makers in middle-income and developing countries to move towards pro-climate resilience policies, codes and standards.

Third, equip businesses with the information and tools they need to integrate climate change considerations into investment decisions. The availability of data on local climate change projections, tools to evaluate the materiality of risks, and information on investment opportunities can in fact help private actors to engage in climate resilience. Governments and DFIs can promote or support investment in the development and dissemination business-relevant data and the provision of business-friendly impact assessment tools. They can also support cost-sharing approaches to facilitate and incentivize climate change risk assessment and support projects structuring.

Fourth, support piloting to demonstrate and create the track record needed to encourage private investment. By demonstrating on-the-ground the financial and technical viability of climate-resilient measures, DFIs can help prove the business case by creating the evidence base needed to encourage private sector's interest in climate resilience.

Fifth, further integrate climate change considerations into the financial system, from DFIs to local financial institutions (banks, microfinance institutions, and leasing companies), insurance companies through to financial regulators. In particular, local banks are in a privileged position to engage businesses to promote the uptake of climate-resilient technologies and practices thereby scaling up climate-relevant investment. This is because they understand local barriers to investments, have local networks, and can engage with widespread and numerous SMEs whose financial needs are typically too small to benefit from direct DFI support.

The Paris Agreement has sent a strong political signal to the financial and investors' communities, calling for a low-carbon and climate-resilient revolution of the global economy. To the investors that have already started the transition, the Agreement has signaled that they can continue to do so with stronger confidence. To the others, it indicates the urgency of changing route. The next step now is in the hands of governments that will have to indicate the way to investors by translating the commitments taken in Paris into adequate national policies and investment plans.

References

African Development Bank (AfDB), Asian Development Bank (ADB), European Investment Bank (EIB), European Bank for Reconstruction and Development (EBRD), Inter-American Development Bank (IADB), International Finance Corporation (IFC) and World Bank. 2012. *Joint Report on Adaptation Finance 2011.* At: http://www.afdb.org/fileadmin/uploads/afdb/Documents/GenericDocuments/Joint%20MDB%20Report%20on%20Adaptation%20Finance%202011.pdf.

African Development Bank (AfDB), Asian Development Bank (ADB), European Investment Bank (EIB), European Bank for Reconstruction and Development (EBRD), Inter-American Development Bank (IADB), International Finance Corporation (IFC) and World Bank. 2013. *Joint Report on Climate Finance 2012.* At: http://www.ebrd.com/downloads/sector/sei/climate-finance-2012.pdf.

African Development Bank (AfDB), Asian Development Bank (ADB), European Investment Bank (EIB), European Bank for Reconstruction and Development (EBRD), Inter-American Development Bank (IADB), International Finance Corporation (IFC) and World Bank. 2014. *Joint Report on Climate*

Finance 2013. At: http://www.eib.org/projects/documents/joint-report-on-mdb-climate-finance-2013.htm.

African Development Bank (AfDB), Asian Development Bank (ADB), European Investment Bank (EIB), European Bank for Reconstruction and Development (EBRD), Inter-American Development Bank (IDB), International Finance Corporation (IFC) and World Bank Group (WBG). 2015. *Joint Report on Multilateral Development Banks' Climate Finance 2014*. World Bank Group. Washington DC. At: http://documents.worldbank.org/curated/en/2015/06/24641149/2014-joint-report-multilateral-development-banks-climate-finance.

Ampri I., A. Falconer, N. Wahyudi, A. Rosenberg, B. Ampera, A. Tuwo, S. Glenday and J. Wilkinson. 2014. *The Landscape of Public Climate Finance in Indonesia*. Indonesian Ministry of Finance and CPI Report. At: http://climatepolicyinitiative.org/publication/landscape-of-public-climate-finance-in-indonesia-3/.

Arnell N. 2011. Policy: The Perils of Doing Nothing. *Nature Climate Change* 1, 193–195, doi:10.1038/nclimate1157. At: http://www.nature.com/nclimate/journal/v1/n4/fig_tab/nclimate1157_F1.html.

Berrang-Ford L., J. D. Ford and A. Paterson. 2011. Are We Adapting to Climate Change? *Global Environmental Change* 21, 25–33.

Bhattarai, R., R. Bogati, N. Bird, M. O'Donnell, J. Lee and E. R. Sigdel. 2011. *Nepal Climate Public Expenditure and Institutional Review*. CPEIR Final Report. Capacity Development for Development Effectiveness Facility for Asia and Pacific. At: http://www.odi.org.uk/resources/docs/7551.pdf.

Brown, J. and N. Kaur. 2009. *Financing Adaptation: Matching Form with Function*. Overseas Development Institute (ODI). At: http://www.odi.org/sites/odi.org.uk/files/odi-assets/publications-opinion-files/5078.pdf.

Brown J., M. Stadelmann, D. Wang, R. Jachnik and T. Kato. 2015. *Estimating Mobilized Private Finance for Adaptation: Exploring Data and Methods*. Climate Policy Initiative in collaboration with the OECD Research Collaborative — Tracking Private Climate Finance. Venice, Italy, At: http://climatepolicyinitiative.org/publication/estimating-mobilized-private-finance-foradaptation-exploring-data-and-methods/

Buchner, B., A. Falconer, M. Hervé-Mignucci, C. Trabacchi and M. Brinkman. 2011a. *The Landscape of Climate Finance*. Climate Policy Initiative (CPI) Report. Venice, Italy. At: http://climatepolicyinitiative.org/europe/publication/the-landscape-of-climate-finance/.

Buchner, B., J. Brown and J. Corfee-Morlot. 2011b. *Monitoring and Tracking Long-Term Finance to Support Climate Action*. Organization for Economic Co-operation and Development (OECD) and IEA. COM/ENV/EPOC/IEA/SLT (2011)3. Paris, France.

Buchner, B., M. Hervé-Mignucci, A. Falconer and C. Trabacchi. 2012. *Global Landscape of Climate Finance 2012*. Climate Policy Initiative (CPI) Report. Venice, Italy. At: http://climatepolicyinitiative.org/publication/global-landscape-of-climate-finance-2012.

Buchner, B., M. Hervé-Mignucci, C. Trabacchi, J. Wilkinson, M. Stadelmann, R. Boyd, F. Mazza, A. Falconer and V. Micale. 2013. *Global Landscape of Climate Finance 2013*. Climate Policy Initiative (CPI) Report. November. Venice, Italy. At: http://climatepolicyinitiative.org/publication/global-landscape-of-climate-finance-2013/.

Buchner, B., M. Stadelmann, J. Wilkinson, F. Mazza, A. Rosenberg, D. Abramskiehn. 2014. *The Global Landscape of Climate Finance 2014*. Climate Policy Initiative (CPI) Report. November. Venice, Italy. At: http://climatepolicy initiative.org/publication/global-landscape-of-climate-finance-2014/.

Buchner, B., C. Trabacchi, F. Mazza, D. Abramskiehn and D. Wang. 2015. *The Global Landscape of Climate Finance 2015*. Climate Policy Initiative (CPI) Report. November. Venice, Italy. At: http://climatepolicyinitiative.org/publication/global-landscape-of-climate-finance-2015/.

Caravani, A., S. Nakhooda and P. Terpstra. 2014. *The Rio Markers in Practice*, Overseas Development Institute (ODI) and World Resources Institute, London, UK and Washington DC.

Clapp, C., J. Ellis, J. Benn and J. Corfee-Morlot. 2012. *Tracking Climate Finance: What and How?* Organization for Economic Co-operation and Development (OECD). COM/ENV/EPOC/IEA/SLT(2012)1. May. Paris, France.

Corfee-Morlot, J., B. Guay and K. M. Larsen. 2009. *Financing for Climate Change Mitigation: Towards a Framework for Measurement, Reporting and Verification*. OECD/IEA Information Paper. October. Paris, France.

Field, C. B. *et al.* (eds.). *Climate Change 2014: Impacts, Adaptation, and Vulunerability Part A: Global and Sectoral Aspects Contribution of Working Group II to the Fifth Assessment Report of the Intergovernmental Panel of Climate Change*. Cambridge: Cambridge University Press.

Hallegatte, S. 2009. Strategic to Adapt to an Uncertain Climate Change. *Global Environmental Change* 19, 240–247.

Höhne, N., S. Khosla, H. Fekete and A. Gilbert. 2012. *Mapping of Green Finance Delivered by IDFC Members in 2011*. IDFC, supported by Ecofys. Frankfurt, Germany. At: http://www.idfc.org/Downloads/Publications/01_green_finance_mappings/IDFC_Green_Finance_Mapping_Report_2012_06-14-12.pdf.

Intergovernmental Panel on Climate Change (IPCC), 2013. Summary for Policymakers. In: *Climate Change 2013: The Physical Science Basis. Contribution of Working Group I to the Fifth Assessment Report of the Intergovernmental Panel on Climate Change* [Stocker, T. F., D. Qin, G.-K. Plattner, M. Tignor, S. K. Allen, J. Boschung, A. Nauels, Y. Xiu, V. Bex and P. M. Midgley (eds.)]. Cambridge University Press, United Kingdom and New York, NY.

International Development Finance Club (IDFC). 2013. *Mapping of Green Finance Delivered by IDFC Members in 2012*. IDFC, supported by Ecofys. Frankfurt, Germany. At: https://www.idfc.org/Our-Program/green-finance-mapping.aspx.

International Development Finance Club (IDFC). 2014. *Mapping of Green Finance Delivered by IDFC Members in 2013*. IDFC, supported by Ecofys.

September. Frankfurt, Germany. At: https://www.idfc.org/Downloads/Publications/01_green_finance_mappings/IDFC_Green_Finance_Mapping_Report_for_2013_12-09-14.pdf.

International Finance Corporation (IFC). 2015a. *IFC Definitions and Metrics for Climate-Related Activities.* Washington D.C.: IFC Climate Business Department. At: http://www.ifc.org/wps/wcm/connect/534495804a803b32b266fb551f5e606b/IFC_Climate_Definitions_2013.pdf?MOD=AJPERES.

Joint-MDB. 2015b, *Common Principles for Climate Change Adaptation Finance Tracking.* African Development Bank (AfDB), the Asian Development Bank (ADB), the European Bank for Reconstruction and Development (EBRD), the European Investment Bank (EIB), the Inter-American Development Bank (IDB) and the World Bank Group. At: http://www.eib.org/attachments/documents/mdb_idfc_adaptation_common_principles_en.pdf.

Jones, L., T. Mitchell, P. S. Villaneuva and S. Standley. 2012. *Coding and Tracking Adaptation Finance: Lessons and Opportunities for Monitoring Adaptation Finance across International and National Scales.* Overseas Development Institute (ODI). September. London, UK. At: https://www.odi.org/sites/odi.org.uk/files/odi-assets/publications-opinion-files/7821.pdf.

Levine, S., E. Ludi and L. Jones. 2011. *Rethinking Support for Adaptive Capacity: The Role for Development Interventions.* Africa Climate Change Resilience Alliance. At: http://community.eldis.org/.5a35bbfb/ACCRA%20Rethinking%20Support%20Report%20Final.pdf.

Machado-Filho, H. 2011. *Financial Mechanisms under the Climate Regime.* In Promoting Compliance in an Evolving Climate Regime, J. Brunnée, M. Doelle and L. Rajamani (eds.), Cambridge University Press, pp. 216–239, (ISBN: 9780521136136).

Michaelowa, A. and K. Michaelowa. 2011. Coding Error or Statistical Embellishment? The Political Economy of Reporting Climate Aid. *World Development* 39, doi:10.1016/j.worlddev.2011.07.020.

OECD DAC. 2013. "Converged Statistical Reporting Directives for the Creditor Reporting System (CRS) and the Annual DAC Questionnaire". Paris: OECD, At: http://www.oecd.org/dac/stats/methodology.htm.

Organization for Economic Co-operation and Development (OECD). 2011. *Handbook on the OECD-DAC Climate Markers.* OECD. Paris, France. At: http://www.oecd.org/dac/stats/48785310.pdf.

Organization for Economic Co-operation and Development (OECD). 2015. *Climate Related Development Finance Project-Level Data* [Internet]; [updated 2015 June 25]. Paris: OECD. At: http://www.oecd.org/dac/stats/climate-change.htm.

Overseas Development Institute (ODI) and Heinrich Böll Foundation (HBF). 2015. *Climate Funds Update Database* [Internet]; [cited July 28]. London, UK and Washington DC: ODI/HBF. At: http://www.climatefundsupdate.org.

Persson, Å., J. T. Klein, C. Kehler Siebert, B. Müller, J. Hoffmaister, M. Lazarus and T. Takama. 2009. *Adaptation Finance under a Copenhagen Agreed Outcome.* Stockholm Environment Institute. October.

Stockholm, Sweden. At: http://www.environmentportal.in/files/adaptation-finance-copenhagen.pdf.

Romani, M. and N. Stern. 2011. *Delivering Climate Finance: Principles and Practice for Funding the Fund.* Grantham Institute. London, UK.

Smit, B. and O. Pilifosova. 2002. *From Adaptation to Adaptive Capacity and Vulnerability Reduction.* In Climate Change, Adaptive Capacity and Development, J. B. Smith, R. J. T. Klein, and S. Hug (eds.), London, UK: Imperial College Press, pp. 9–28.

Terpstra, P. 2013. *The Difficulty of Defining Adaptation Finance.* World Resources Institute, October 2013, at: http://www.wri.org/blog/difficulty-defining-adaptation-finance.

Terpstra, P., A. Peterson Carvalho and E. Wilkinson. 2013. *The Plumbing of Adaptation Finance: Accountability, Transparency and Accessibility at the Local Level.* World Resources Institute Working Paper, November. Washington DC. At: http://www.wri.org/sites/default/files/adaptation-finance-plumbing.pdf.

Tirpak, D., A. Ronquillo-Ballesteros, K. Stasio and H. McGray. 2010. *Guidelines for Reporting Information on Climate Finance.* WRI Working Paper. World Resources Institute. Washington DC. At: http://pdf.wri.org/working_papers/guidelines_for_reporting_information_on_climate_finance.pdf.

Tirpak, D., K. Stasio and L. Tawney. 2012. *Monitoring the Receipt of International Climate Finance by Developing Countries.* World Resources Institute. Washington DC. At: http://pdf.wri.org/monitoring_receipt_of_international_climate_finance_by_developing_countries.pdf.

Trabacchi, C. and F. Mazza. 2015. *Emerging Solutions to Drive Private Investment in Climate Resilience.* Climate Policy Initiative (CPI). June. Venice, Italy. At: http://climatepolicyinitiative.org/publication/emerging-solutions-to-drive-private-investment-in-climate-resilience/.

United Nations Development Programme (UNDP). 2013. *Country Reports on Climate Public Expenditures and Institutional Reviews* (CPEIR). At: http://www.climatefinance-developmenteffectiveness.org/publications.html.

United Nations Environmental Programme DTU Partnership (UNEP DTU). 2015. [Cited as AGR, 2015]. The Adaptation Finance Gap Report, with insights from the INDCs. At: http://web.unep.org/adaptationapreport/content/adaptation-gap-reports.

United Nations Framework Convention on Climate Change (UNFCCC). 1992. *United Nations Framework Convention on Climate Change.* United Nations. FCCC/INFORMAL/84, GE.05-62220 (E) 200705. At: http://unfccc.int/resource/docs/convkp/conveng.pdf.

United Nations Framework Convention on Climate Change (UNFCCC). 2010. *The Cancun Agreements: Outcome of the Work of the Ad Hoc Working Group on Long-Term Cooperative Action under the Convention* (1/CP.16). UNFCCC.

United Nations Framework Convention on Climate Change (UNFCCC). 2011. *Report of the Conference of the Parties on its Seventeenth Session, Held in*

Durban from 28 November to 11 December 2011. Decisions adopted by the Conference of the Parties. Decision 2/CP.17. FCCC/CP/2011/9/Add.1. At: http://unfccc.int/resource/docs/2011/cop17/eng/09a01.pdf.

United Nations Framework Convention on Climate Change Standing Committee on Finance (UNFCCC SCF). 2014. "Biennial Assessment and Overview of Climate Finance Flows". At: http://unfccc.int/cooperation_and_suppport/financial_mechanism/standing_committee/items/8034.php.

Vivid Economics. 2015. Building on Evidence Base on Private Sector Engagement in Financing Climate Change Adaptation. Available at: http://www.ebred.com/documents/climate-finance/building-on-evidence-base-on-private-sector-engagement-in-financing-climate-change-adaptation.pdf

World Bank (WB). 2012. Turn Down the Heat: Why a 4°C Warmer World Must Be Avoided. Washington DC: World Bank. http://documents.worldbank.org/curated/en/865571468149107611/Turn-down-the-heat-why-a-4-c-warmer-world-must-be-avoided

World Bank Group *et al.* (WBG). 2015a, "Common Priniciples for Climate Mitigation Finance Tracking". Jointly Developed by the Climate Finance Group of Multilateral Development Banks (MDBs) and the International Development Finance Club (IDFC). April 2015. Washington DC. At: http://www.worldbank.org/content/dam/Worldbank/document/climate/common-principles-for-climate-mitigation-finance-tracking.pdf

Chapter 4

Engaging Private Sector and Mobilizing Private Finance through Mitigation Actions in Developing Countries

Alina Averchenkova

Grantham Research Institute on Climate Change and the Environment at the London School of Economics, Clements Inn Passage, London WC2A 2AZ
a.averchenkova@lse.ac.uk

Abstract

Over the past several years, there has been a significant shift in the recognition of the role of the private sector in financing low carbon transition in developing countries. While the volume and variety of climate-financing resources are growing, more work is required on the part of the public sector to engage private sector and catalyze private financing at scale. Low carbon development strategies and national mitigation actions can become the vehicles for the redirection of investment through creation of policy frameworks that remove barriers and create incentives for low carbon investment. This chapter provides an overview of the key issues in relation to engaging the private sector in mitigation actions in developing countries. It discusses the importance of the private sector for the low carbon transition, the various types of private sector players and their motivations vis-à-vis climate change mitigation, as well as barriers and opportunities for private investment. It further identifies opportunities for engaging private sector in mitigation actions and highlights some of the recent practical experiences with the implementation of the nationally appropriate mitigation actions (NAMAs). The analysis of practical experience shows that there is great interest in engaging the private sector in the development and implementation of mitigation actions. Yet it is still to become a primary focus in practice in some developing countries, particularly those where mitigation actions are at earlier stages.

1. Introduction

Over the past several years, there has been a significant shift in the recognition of the role of the private sector in implementing and financing climate change mitigation in developing countries. In the past in the international negotiations, the issue of private finance has often been seen as a subject brought up to divert attention from the need to provide public finance by developed countries and to pass on the responsibility to the private sector. Domestically, private sector has been primarily thought of a risk factor and potential barrier to the implementation of climate change mitigation programmes and policies. Yet an increasing number of studies have assessed the scale of the global financing challenge to enable the transition to low carbon economy and highlighted the critical role of the current and future private investment in this context. At the same time, more practical experience is being generated on the implementation of mitigation actions, demonstrating that engagement of the relevant private sector actors in mitigation projects and programs early on allows avoiding some of the pitfalls in the design and implementation. It also enables mobilization of private sector resources at scale.

The outcomes of the Paris summit in December 2015 demonstrated a strong recognition that private sector is an essential link in the chain of the elements that need to be connected to ensure transition to low carbon economy and needs to be brought aboard and becomes a partner in the process of designing and implementing climate change mitigation.

While the understanding of the importance of these issues has grown, the practical experiences of successfully engaging the private sector and mobilizing private finance through mitigation programmes are still a relatively new area, with more information being constantly demanded by the climate practitioners. This chapter provides an overview of the key issues in relation to engaging the private sector in mitigation actions in developing countries. It discusses the importance of the private sector for the low carbon transition, the various types of private sector players and their motivations vis-à-vis climate change mitigation, as well as barriers and opportunities for private investment. It further identifies opportunities for engaging private sector in mitigation actions and highlights some of the recent practical experiences with the implementation of the nationally appropriate mitigation actions (NAMAs).

2. Defining Private Sector

The importance of the private sector in the transition to low-emission development is hard to underestimate. Some private sector actors provide technology and finance for projects, while others implement the projects and deliver relevant services. Private sector is not homogenous. It includes various actors, all of whom have different economic interests and modes of operation, as well as different roles to play in the context of climate change mitigation. Some may have opportunities to expand their business into new technologies, products or markets. Others may stand to lose under their current business models due to the lower competitiveness of high carbon technologies in the future. Understanding the key players on both ends of the spectrum, the impacts they are likely to face and working out solutions for the transition helps policy makers to get greater buy-in for the proposed policies and identify more effective ways to deploy limited public resources to mobilize additional private finance.

First, a distinction can be made between financiers, project developers and owners, and service companies. Corporate project owners and technology providers are essentially the entities developing project ideas and implementing them, provided that they are able to raise the necessary capital. These actors are often the primary targets of low carbon policies and mitigation actions aimed at creating regulation, such as technology standards, carbon taxes and emission-trading schemes. Financiers include private sector actors that provide debt and equity financing for commercial investment to project owners and technology providers. Here a distinction should be made between venture capital investors, institutional investors and retail lenders. Venture capital investors are likely to assume higher risk in the early stages of a project or when a new technology is being applied, but they expect higher returns on their investment. Institutional investors are prepared to see lower returns, but make only relatively secure investments. Retail lenders (e.g., commercial banks) fall between these two. Finally, service providers (often consulting companies) are involved in the development of the underlying mitigation projects and policies, measuring, reporting and verifying (MRV) emissions, carbon trading, etc.

Further distinction can be made between companies that are the primary targets of mitigation programmes and those that belong to the related value chain. Examples of the former include renewable energy developers who might directly benefit from feed-in-tariffs or car manufacturers who may receive a subsidy or tax break for producing or selling low-emission vehicles. Examples of the latter include producers of equipment such as wind turbines or solar panels, or suppliers of batteries for electric vehicles. Assessment of the role of the private sector in the low carbon transition and of the ways to scale-up private finance requires consideration of the entire value chains. This is necessary in order to understand the full spectrum of potential barriers to the development of such a market. For example, the availability of replacement parts and repair service for electric vehicles in the local market may affect the feasibility of implementing this technology, even if competitively priced electric vehicles themselves are available. Similarly, when considering private-sector financiers, the complete spectrum of potential investors should be identified and analyzed.

3. The Role of Private Sector in Low Carbon Transition

According to the Climate Policy Initiative (CPI), private actors are the largest source of global climate finance to date, and invested

USD 193 billion or 58% of total climate finance flows in 2013. Most of it came from project developers and corporate actors and manufacturers [Buchner *et al.* 2014]. There had been some decrease in the amount of private climate finance in 2013 compared to 2012, which according to CPI, was mainly due to the falling cost of certain renewable energy technologies, e.g., of photovoltaic system (solar PV). To achieve the same level of solar deployment in 2013, it cost USD 40 billion less than it did in 2012.

However, the level of private climate finance has to grow significantly in the future to keep global temperature increase below 2°C. The New Climate Economy report [NCE, 2014] estimates that around USD 90 trillion needs to be invested in infrastructure assets up to 2030 to meet the demand of the growing population globally for energy, water, food and other basic services. These investments will be critical to whether or not the world locks itself into a high- or low-carbon growth trajectory over the next 15 years [Bhattacharya *et al.*, 2015]. According to the NCE [2014], in order to ensure that these investment lead to low carbon transition, an additional USD 4.1 trillion would need to be invested over this period, with a large share of this investment coming from the private sources.

The Intergovernmental Panel on Climate Change (IPCC) [Gupta *et al.*, 2014] estimates that the global total annual investment required in the energy sector under the scenarios compatible with limiting temperature rise to 2°C is about USD 1,200 billion up to 2030. These assessments point to a decline in annual investment in fossil-fired power plants without carbon dioxide capture and storage (CCS) by USD 30 billion, and to an increase in investment in low-emissions energy technologies by around USD 147 billion per year and in energy efficiency by USD 336 billion in the period 2010–2029 respectively.

In order to better understand how private investment at such scale can be mobilized for climate change mitigation, it is useful to consider the motivation of private sector players, key determinants of their investment decisions, as well as potential governmental intervention that can help enable the actions (see Table 1). It is also

Table 1. Key Motivations and Determinants of Corporate Investment in Mitigation.

Motivations	**Drivers**	**Determinants**	**Potential governmental intervention**
Profit-making	— Expectation of positive return — Attractive pay-back period for investment compared to alternative options — Access to new markets	— Risk-return relationship for the investment — Cost of capital (finance) — Liquidity (for investors) — Capital and operational expense required — Access to technology and skills; technological change — Security/ stability of the market and policy frameworks	— Policy instruments to balance risk-return (either through pricing full carbon externalities or setting incentives for low carbon options), e.g., carbon tax, ETS, feed-in-tariffs, tax credits, etc. — Financial instruments transferring part of the investment risk to the public sector — Capacity building and opportunities for partnerships
Risk management	— Preparing for future regulation — Fiduciary obligations — Investor pressure	— Likelihood of policy being introduced — Exposure under the proposed policy change	— Consultation with the private sector on policy proposals — Encouragement and assistance to voluntary GHG measurement and management programmes

(*Continued*)

Table 1. (*Continued*)

Motivations	Drivers	Determinants	Potential governmental intervention
			— Requirements for information disclosure/reporting — Capacity building
Reputation and license to operate	— Desire to improve image — Investor, consumer, NGO pressure — Ethical considerations	— Internal leadership, knowledge and capacity to run CSR programmes — Drive from the headquarters or from up the value chain (especially for the MNCs)	— Rewarding champions — Showcase work — Financial incentives for best performers (e.g., tax incentives or access to public resources)

useful to note that private sector investment into mitigation may take different forms (see Table 2). The latter is adapted from an empirical study that analyzed motivations and responses of the FTSE 100 group of companies (100 largest companies listed on the London Stock Exchange) to climate change [Okereke, 2007].

Comprising the largest group of climate-change investors globally, the private sector is mainly driven by profit-making and favorable returns on investments. In this context, any project investment decisions are made based on the relationship between risk and return. According to Buchner *et al.* [2014], three-quarters of climate finance flows attributed to the private sector in 2013 were invested with the expectation of earning commercial returns. Yet, low carbon technologies are often not competitive locally against their high carbon alternatives, in particular in developing countries, due to a number of barriers affecting risk-return relationship [Waissbein *et al.*, 2013]. Most of these have to do with the characteristics of the investment

Table 2. Types of Mitigation Activities by FTSE 100 Companies.

Type of actions	Examples of company responses
Basic technological change	Instalment of new or replacement of existing office equipment and machinery with the bid to curtail in-house GHG emissions or increase energy efficiency (low energy bulbs; refurbishing property; replacement of boilers or generators).
Behavioral change	Change in the behavior of employees to reduce GHG emissions (double-sided printing; reducing printing rate; switching off lights; separating and recycling waste; videoconferencing, etc.).
Educating the public and suppliers	A growing number of companies among the FTSE group show active interest in educating the public and the suppliers on climate change.
Participation in emission trading (ETS) and offsetting schemes	This mainly applies to the companies operating in geographies with limitations on GHG emissions, as well as to voluntary initiatives for offset emissions as part of CSR programmes.
Finding new use(s) or market for existing products	Explore the possibilities of developing alternative or multiple uses of their CO_2 products.
Developing low carbon products in response to emerging demand	Satisfy the need for new and existing low carbon products and technologies to replace high carbon ones.
Investment in low carbon portfolios	Investments in low carbon and process related technology, such as sustainable energy like wind farm, solar, biomass, and water; or waste management or recycling plants and schemes.
Acquisition of assets that balance companies' carbon portfolio	Companies are increasingly considering the extent to which an acquisition affects their carbon profile and/or disinvesting from carbon intensive industries.
Fundamental technological shift and innovations	"Game changing" deep transformation in the focus or method of operation, e.g., changes in mode of manufacturing, redesign of process machines; putting waste to good use. Examples include bio-fuels, use of biomass feedstock, carbon capture and storage, and integrated gasification combined cycle (IGCC) technology.

Source: Adapted from Okereke [2007].

itself or with the country's wider investment climate. These barriers vary by sector, technology and country and depend on the overall economic and political environment in each, as elaborated below. Hence, the key to mobilizing private sector investment driven by the returns on investment is a combination of better and more stable policies at the country level, more effective project planning, and appropriate risk-sharing instruments, which can all help to lengthen investor time horizons [Bhattacharya *et al.* 2015].

Other factors that motivate private sector actions and investment into climate change mitigation include risk management based on the expectation of a future policy change (e.g., introduction of an emission trading scheme (ETS) or carbon tax, or new vehicle emission standards), as well as reputation and licence to operate in the local community. The latter may also be influenced by the overall Corporate Social Responsibility (CSR) policy, in particular in the case of multinational companies (MNCs), or by the requirements from upstream in the value chain (e.g., from the project developer and investor to the local suppliers). While these factors seem to play a more modest role than profit-making in mobilizing private climate investment, as suggested by the above data by the CPI, they can still be an important driver in engaging private sector. This is particularly true in developing countries at the earlier stages of implementation of their low carbon strategies. Governments can help mobilize actions by private sector in this context through consultation with the private sector on policy proposals to make them aware of the upcoming policy change; encouraging and providing capacity building for voluntary greenhouse gas (GHG) management programs; launching reporting initiatives and setting requirements for disclosure; and showcasing efforts by the private sector and rewarding champions through recognition or other incentives (e.g., tax incentives or access to public resources).

4. Tools to Address Barriers to Private Finance

Given the central importance of the risk-return balance, as discussed above, in order to attract private sector investment to finance low carbon transition a country needs to provide potentially high return rates or to lower the risks to investors. Hence the main challenge is not generating capital, but redirecting it [Waissbein *et al.*, 2013]. This requires that a number of barriers that make low carbon investment more expensive to be addressed. The Climate and Development Knowledge Network (CDKN) guide to financial barriers for climate investment makes a distinction between several types of barriers:

- the barriers related to the overall financial, regulatory and political climate in the country and the size of the market;
- the barriers related to general competitiveness issues of low carbon options in the environments where externalities are not included in the absence of price on carbon; and
- the barriers related to the lack of experience with new and unproven technologies, and lack of information leading to behavioral failures [Berliner *et al.*, 2013].

Companies operating in developing countries often face a number of difficulties in obtaining finance for projects due to overall national political and currency risks and failures in the local financial markets. This normally means higher interest rates and shorter lending periods than in the more established and stable markets. In addition to these overall challenges, low carbon investment is affected by further financial barriers, such as high upfront costs, small project sizes, split incentives between owners and users (e.g., between landlords and tenants for energy efficient buildings) and misallocation of resources through fossil fuel subsidies [GIZ, 2013]. Other barriers include limited experience with and lack of technical or other capacity and information on low carbon technologies, which often keeps financiers from lending to low carbon projects or results in higher interest rates.

A number of public sector interventions have been applied to improve attractiveness of low carbon investment (see Table 1). The main objective of those interventions is to balance risk-reward relationship so that low carbon technologies can compete locally with higher carbon alternatives and therefore become attractive for the private sector. This can be done through policy instruments, financial instruments, and capacity building, or the combination of the all of the above.

Policy instruments aim at improving the relative return on low carbon investment either through pricing carbon externalities (e.g., through carbon tax or an ETS) or through providing incentives for low carbon technologies (e.g., through feed-in-tariffs, tax credits, etc.). For example, feed-in-tariffs or power purchase agreements increase the size and improve security of the expected return from renewable energy projects. Similarly, a functioning carbon market can play a similar role.

Financial instruments seek to transfer the investor risks to public actors (e.g., to development banks), for example through loan guarantees, political risk insurance and public equity co-investments [Waissbein *et al.*, 2013]. By transferring risk against project failure these instruments lower the financing costs (debt and equity), which can significantly affect the competitiveness of low carbon projects in developing countries. In situations in which local markets are not large enough to attract investors, the aggregation of projects into a portfolio allows for risk sharing and economies of scale. A recent study analyzed several case studies of low-carbon investment, focusing specifically on mobilizing investment from the private sector [Buchner *et al.*, 2012]. This research showed that lessening risk is particularly important for enabling private investment.

Risk transfer and sharing instruments have been widely used in other development areas, but are relatively new to climate-related investment. They are particularly attractive given that no cash expenditure is required upfront and public funding is only used in the case of a guarantee being called in (when a project has major problems or fails). According to a recent study [Micale *et al.*, 2013], multilateral development banks provide risk coverage to sectors and countries for which private providers either charge very high rates or would not consider at all. Most of these institutions cover a range of policy-related risks, while International Finance Corporation (IFC) and International Bank for Reconstruction and Development (IBRD) cover the full range of risks, including commercial ones.

Finally, capacity building and sharing information on low carbon technologies and their performance with finance providers and project developers also help in removing the related barriers mentioned earlier. The instruments mentioned can be combined in order to mobilize different types of investment from various kinds of investors so that funds cascade down to the level of technical implementation [GIZ, 2013].

These tools are currently being designed and implemented in the context of low emission development strategies (LEDs) and NAMAs in many developing countries around the world.

5. NAMAs as a Vehicle to Mobilize Finance

NAMAs as a concept emerged in the Bali Action Plan negotiated in 2007 [UNFCCC, 2008] and then further developed through the Copenhagen Accord and the Cancun Agreements, as well as the subsequent decisions adopted by the 18th Conference of the Parties to the UNFCCC in Doha in 2012. These decisions established that developing countries will undertake NAMAs voluntarily in the context of sustainable development and that such actions will be supported and enabled by technology, financing, and capacity building, and invited countries are to communicate NAMAs aimed at achieving deviation from business as usual emissions.

However, there is not a clear definition of NAMAs provided in these internationally agreed documents. Much of the interpretation of what they mean is provided by think tanks, non-governmental organisations (NGOs) and international agencies working on mitigation actions on the ground [Coetzee *et al.*, 2014]. According to the UNFCCC secretariat's website, "NAMAs refer to any action that reduces emissions in developing countries and is prepared under the umbrella of a national governmental initiative [...] and can be policies directed at transformational change within an economic sector, or actions across sectors for a broader national focus" [UNFCCC, 2015]. There is an understanding, however, that NAMAs should aim at achieving a reduction in emissions relative to "business as usual" emissions in 2020.

NAMAs offer an opportunity to address the barriers to low carbon development and to mobilize investment through putting in place policy frameworks and incentives for private investment. Essentially the objective of such policy interventions is to identify the potential for GHG emission reduction in key sectors, determine the technological and financial barriers to lower emission alternatives and to design policies and regulations, and put in place incentives and other measures to address these barriers. Early engagement with the private sector around the development of NAMAs therefore becomes crucial for success. Going forward, the experiences with implementing NAMAs are expected to be scaled-up to help achieve more ambitious emission reduction objectives that countries have been setting in their intended nationally determined contributions (INDCs) pledges for the periods up to 2025–2030 ahead of the Paris climate summit that took place in December 2015.

As of 1 October 2015, there were 99 NAMAs registered in the UNFCCC pipeline, communicated by 57 countries as well as the African Group, with the largest number coming from Latin America (see Figure 1). These projects are split between those submitted for recognition only and therefore being implemented by developing countries with the use of their own resources, and those that are either seeking support for preparation of a detailed NAMA plan or for its implementation. Africa is the region with the largest share of NAMAs seeking support for preparation phase, while Europe and Central Asia have the greater share of projects needing support for implementation. Most NAMAs submitted to the UNFCCC secretariat are focused on energy efficiency, renewable energy and transport. These three sectors capture two-thirds of NAMAs (see Figure 2).

6. Experience of Engaging Private Sector through NAMAs to Date

Analysis of the initial experience of the development and implementation of NAMAs in developing countries shows that to date there has been some experience of engaging the private sector in

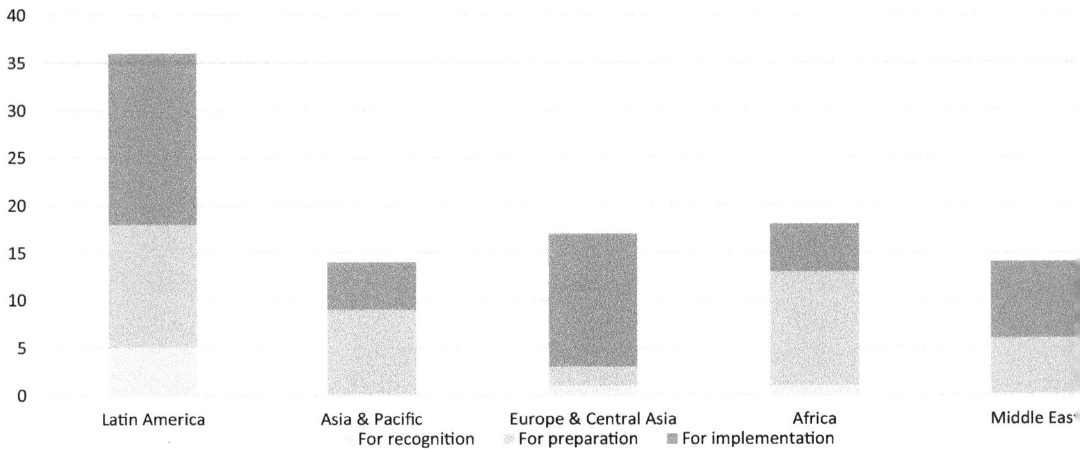

Figure 1. Overview of NAMAs Submitted to the UNFCCC Secretariat as of 1 October 2015.

Source: Calculated by author based on UNEP DTU NAMA Pipeline Analysis and Database [2015]. Available at http://www.namapipeline.org. Accessed on 1 October 2015.

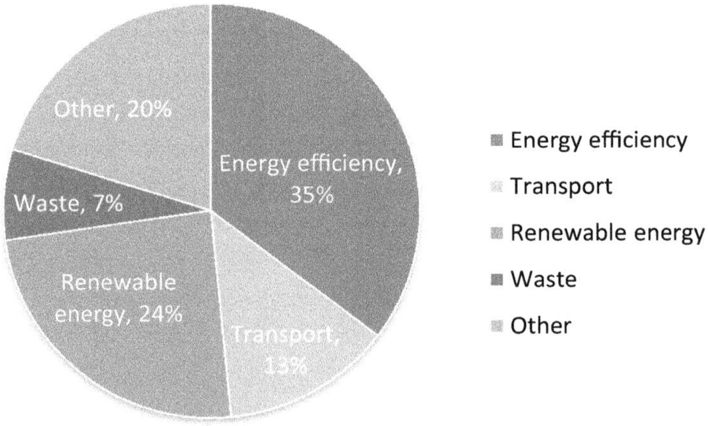

Figure 2. UNFCCC Global NAMA Pipeline by Sector as of 1 October 2015.

Source: Calculated by author based on UNEP DTU NAMA Pipeline Analysis and Database [2015]. Available at http://www.namapipeline.org. Accessed on 1 October 2015.

mitigation activities, yet there is limited direct evidence of private finance already specifically generated through NAMAs. The lack of direct evidence on the private sector finance mobilized is due to the fact that NAMAs are relatively new with very few having reached implementation stage so far.

As of 8 November 2015, there were only seven NAMAs registered in the UNFCCC registry for recognition, meaning they were not requesting support for implementation. These NAMAs are generally at a more advanced stage than those seeking support. While the information provided in the registry does not allow to assess the scale of private investment to be mobilized as a result of these NAMAs, it is possible to evaluate the types of interventions and instruments that are being used in these programmes to induce investment (see Table 3).

Most of these NAMAs include either public procurement of low carbon investment (e.g., through tenders for renewable energy installations) and/or some form of regulation or standard (e.g., efficiency standards for buildings). There are examples of using either incentives (tax benefits) or voluntary partnerships to build capacity and encourage management of GHG emissions (Clean Production Agreements (CPA)). Yet, none of these NAMAs are focusing explicitly on financial instruments for risk sharing and transfer.

Examination of a larger sample of NAMA programs that are still at earlier preparatory stages allows to draw further insights into developments on the ground. A recent study [Averchenkova, 2014] assessed the experiences within the United Nations Development Programme's (UNDP) Low Emission Capacity Building Programme (LECB) launched in 2011 to support 25 developing countries to build the public- and private-sector capacities needed to scale up mitigation actions through LEDs and NAMAs. Through a survey and interviews of public and private sector representatives engaged in the programme,[1] the study identified that the primary challenges

[1]The survey was disseminated to policymakers and national experts in all countries participating in the LECB. In all, 31 responses to the survey have been received, covering 17 countries: Bhutan, Chile, Costa Rica, Democratic Republic

Table 3. Examples of Instruments Used in the NAMAs under Implementation Submitted to UNFCCC for Recognition.

Type of intervention	Country NAMA example	Instruments used
Policy Instruments		
Incentives	Uruguay NAMA for recognition: Promotion of renewable energy in the primary energy mix.	Tax benefits for renewable energy projects; Power purchase agreements for biogas.
Regulation and pricing mechanisms	Uruguay NAMA for recognition: Promotion of renewable energy in the primary energy mix.	Regulation for solar thermal power utilization in buildings.
	Colombia NAMA for recognition: Integrated improvement of road-based freight sector.	Freight logistics policy, a freight improvement program and freight vehicle scrappage and fleet renewal.
	Serbia NAMA for recognition: Construction of new energy efficient buildings based on energy efficiency regulation.	Regulations on energy efficiency in building for all new buildings, e.g., better thermal insulation to reduce heat energy consumption.
Partnerships and Capacity Building		
Public procurement and PPPs (contractual basis)	Uruguay NAMA for recognition: Promotion of renewable energy in the primary energy mix.	Tenders for wind and solar PV installations by private sector.
	Uruguay NAMA for recognition: Expansion of electricity generation from sustainable forestry biomass by-products.	Tenders to install forestry biomass power to deliver electricity to the national grid.
	Ethiopia NAMA for recognition: Construction of the Light Rail Transit system and a modal shift of freight transport from road to rail network powered through renewable energy.	Financed by the government based on borrowed capital.

(*Continued*)

Table 3. (*Continued*)

Type of intervention	Country NAMA example	Instruments used
	Uruguay NAMA for recognition: LNG Terminal with regasification capacity.	Financing of an installation of an LNG.
Voluntary partnerships	Chile NAMA for recognition: The Clean Production Agreements (CPA).	Voluntary agreements between the companies and the government to implement clean production based on specific goals and actions according to the best available techniques in the market. 70% funded by the government and 30% by the private sector. Public funding does not support acquisition of technology.

Source: Based on the information in the UNFCCC NAMA registry [2015]. Available at http://www4.unfccc.int/sites/nama/SitePages/Home.aspx. Accessed on 8 November 2015.

in engaging private sector relate to the lack of clear understanding among policy makers of what the role of the private sector should be in the context of NAMA development and implementation, and the lack of clarity on the types of players that should be involved and the uncertainty as to when and how to go about it.

In response to the survey's multiple choice questions on the reasons for engaging the private sector, over 80% of the respondents

of Congo (DRC), Ecuador, Egypt, Ghana, Kenya, Lebanon, Malaysia, Mexico, Moldova, Peru, Tanzania, Uganda, Vietnam and Zambia. The survey was complemented by desk reviews and 20 interviews in 2013 as well as further eight interviews in 2014–2015, the later focusing specifically on Africa and the Middle East.

indicated that private-sector involvement is critical in terms of financing the implementation of the mitigation programmes, and over 70% indicated that it is important to seek feedback from the private sector on proposals for such programmes or projects. At the same time, only about one-third of the respondents felt that getting the private sector on board will help secure political support, while around 7% expected that there will be no private-sector interest in the issue in their country. Some experts through additional open comments noted that engagement of the private sector will help build the capacities and technical expertise of teams working on the design of GHG mitigation programmes and help clarify national priorities.

In terms of practical steps to engage private sector, there are some positive examples on the first steps being taken from the Africa and the Middle East initiatives, many of which are at the earlier stages than in other regions (see Table 4). These include initiative to educate and train the private sector to build its capacity to identify, develop and implement mitigation actions (e.g., in Ghana, Egypt and Kenya). There are also efforts being made to inform private sector and to consult on the NAMA ideas and proposals (e.g., Lebanon, Morocco, Zambia and DRC). Some countries are implementing voluntary mitigation partnerships, as the first step towards engaging carbon-intensive businesses in mitigation actions (e.g., Uganda and Lebanon).

A number of countries in other regions have made private sector the primary target and a partner of their NAMAs (e.g., Mexico). In some countries, there is interest from the private sector in potentially undertaking bottom-up initiatives within the context of NAMAs (e.g., in Chile).

Overall, the primary target of the effort to engage private sector so far has been companies operating in the sectors affected by a NAMA. Other private-sector actors who have been engaged in NAMA development include consulting companies, Clean Development Mechanism (CDM) project developers and traders and in some cases local commercial banks. Yet, there has been little engagement of the financing and investor communities to date [Averchenkova, 2014].

Table 4. Examples of Engaging Private Sector in the NAMAs under the LECB in Africa and Middle East Region.

Type of intervention	Country NAMA example and instruments used
Voluntary partnerships	Uganda: Workshop with over 100 private sector representatives and stakeholders on opportunities for private sector engagement and financing in mitigation activities. Private sector representatives signed a declaration committing to reduce carbon emissions.
	Lebanon: Voluntary GHG Reporting initiative to benchmark companies through measuring and reporting their emissions. The project stimulates capacity building and development of the MRV for GHGs in the private sector.
Information sharing and consultation	Morocco: Consultation with private sector on the NAMA proposals.
	Zambia: Consultation with the private sector on project ideas; inclusion into the NAMA working group; selection of partners.
	Democratic Republic of Congo: Started with asking the representatives of business about the barriers they face, their experiences in implementing climate change activities and with engaging with public sector in the past, as well as about the opportunities they see to work together on NAMAs in agriculture, energy efficiency and waste management.
	Lebanon: Ran an extensive period of consultation with the stakeholders, including private sector, to generate ideas for NAMA projects, which were subsequently prioritized and a shortlist put forward for development.
Education and training of the private sector	Egypt: MRV system for two fertilizer factories and two iron and steel factories.
	Ghana: Prepared an Investor Guide for NAMAs for the private sector. Outreach on climate change to Club 100 companies that have Corporate and Social Responsibility budgets. Work with the Ghana Investment Promotion Centre to engage private sector players in NAMAs generating ideas, packaging them and taking them to investors.
	Kenya: Creation and work of specialized centers to promote private sector investment and clean technology incubation.

Source: Based on the interviews with the local LECB staff in the respective countries.

A frequent challenge is that ministries of environment have limited contacts with decision makers in the companies. Therefore, often sectorial ministries such as energy, transport, industry, and agriculture are better suited to reach out given their regular contact with companies operating in their sectors. In addition, in some countries where the private sector was not engaged early on, there is now industry opposition to NAMA proposals. Most private-sector engagement to date has therefore been done through local business associations, as there are usually established relationships present and they are convenient as a single point of contact. The downside is that it limits the outreach to the active members of the associations and may exclude some sectors. Furthermore, representatives in the associations normally come from environment and safety departments, with limited power to influence the investment decisions. In the future, it is important to extend outreach to potential champions in the relevant sectors and to non-environmental business departments.

7. Mobilization of Private Finance through Public Support from Developed Countries Globally

While NAMAs are still at early stages of development and implementation, there are many other climate change mitigation projects and initiatives being implemented through various domestic, bilateral and multilateral climate finance channels globally. A recent analysis by the Organisation for Economic Co-operation and Development (OECD) in cooperation with the CPI assessed the amount of public finance and direct co-financing by the private sector mobilized for climate change in 2013–2014 [OECD, 2015]. This is done in the context of assessing the progress towards the commitment made by developed countries to mobilize USD 100 billion annually for assisting climate change mitigation and adaptation efforts in developing countries by 2020.

According to this analysis, total finance mobilized was USD 62 billion in 2014, up from USD 52 billion in 2013, making it an average of USD 57 billion annually over the 2013–14 period [OECD, 2015]. Over three-quarters of this amount was to support mitigation activities. The data includes finance provided by donor governments through various instruments and institutions, including non-concessional loans. It also includes private funding for climate-related projects that has been directly mobilized by developed country public financial interventions through co-financing. Preliminary estimate of the private co-financing mobilized shows that it stands at around 25% of the total finance mobilized. Importantly, as the authors point out, this data does not capture the indirect mobilization of private sector finance due to capacity building, budgetary support and improved domestic policy frameworks discussed earlier.

There are also some concrete examples on the mobilization of private sector finance through particular financial instruments. For example, GEF climate change mitigation projects on average mobilized USD 1.3 from the private sector per USD 1 GEF grant in 2013–14 [GEF, 2015]. This is double of the figure for the GEF portfolio on average as a whole. However, recently in the projects using innovative "non-grant instruments" financial instruments mobilization rates from the private sector have reached USD 6.8 for each USD 1 in GEF grant. Such projects, for example, use GEF funds for first loss risk guarantees, or junior debt and equity positions in larger climate finance packages. This example confirms that there is significant potential to increase mobilization of private-sector finance for mitigation going forward.

8. Conclusions

The case for mobilizing private financing for low-carbon technology at scale over the next couple of decades — much of it in developing countries — is clear. While the volume and variety of climate-financing resources are growing, more work is required on the part of the public sector to engage private sector and catalyze private financing at the level required to meet the climate change challenge. Low carbon development strategies, NAMAs and in the future the INDCs should increasingly become the vehicles for the redirection of investment through creation of policy frameworks that remove barriers and create incentives for low carbon investment.

In order to attract an active engagement and investment from the private sector, such policy frameworks will need to provide long-term clarity and stability for the investor. To do so, they should introduce measures to reduce risk for investments in the short term, lowering the cost of capital or making access to it easier, or enhance the competitiveness of low carbon technologies by pricing carbon externalities or introducing regulations and standards. Capacity building of private-sector actors may also help facilitate investment and engagement.

The analysis of practical experience shows that there is great interest in engaging the private sector in the development and implementation of mitigation actions. Yet, it is still to become a primary focus in practice in some developing countries, particularly those where mitigation actions are at earlier stages.

Going forward, achieving the objectives set by countries in their INDCs for 2025–2030 will require an improvement in the global energy intensity by 2030 at rate almost three times faster than the rate seen since 2000, with 70% of additional electricity generation in the power sector being low carbon [IEA, 2015]. Clearly, this scale of transition will require major transformation and investment from the private sector. At the same time, the economy-wide and/or sectoral quantifiable goals set by most countries in their INDCs provide a more solid basis to mobilize private investment than was previously feasible under largely project-based NAMAs.

In order to ensure these objectives are achieved, it is important that the international frameworks for mitigation actions, finance and technology cooperation are sufficiently flexible to allow for a variety of policy solutions and incentives at the national level depending on the circumstances of particular countries and sectors. Financial institutions, such as the Green Climate Fund and its Private Sector Facility, should put leveraging private-sector finance through public funding as one of their priorities. In this context, building capacity of the private sector, in particular in the financial and investment community in developing countries, among potential project developers and technology providers, will be very important.

References

Averchenkova, A. 2014. Mobilising Private-Sector Engagement in LEDS and NAMAs, United Nations Development Programme Low Emission Capacity Building Programme (UNDP LECB).

Berliner, J., C. Grüning, K. Kempa, C. Menzel, and U. Moslener. 2013. *Addressing the Barriers to Climate Investment.* Climate and Development Knowledge Network (CDKN).

Bhattacharya A., J. Oppenheim and N. Stern. 2015. Driving Sustainable Development through Better Infrastructure: Key Elements of a Transformation Programme. Global Economy & Development Working Paper 91.

Buchner, B., T. Heller, and J. Wilkinson. 2012. Effective Green Financing: What Have We Learned So Far? Climate Policy Initiative (CPI).

Buchner B., M. Stadelmann, J. Wilkinson, F. Mazza, A. Rosenberg and D. Abramskiehn. 2014. *Global Landscape of Climate Finance 2014.* Climate Policy Initiative (CPI).

Coetzee, K. and H. Winkler. 2014. The International Policy Context for Mitigation Actions, *Climate and Development* 6.1, 4–11.

Deutsche Gesellschaft für Internationale Zusammenarbeit (GIZ). 2013. *NAMA Tool.* At: https://unfccc.int/files/cooperation_support/nama/application/pdf/namatool.pdf.

Global Environment Facility (GEF). 2015. Mobilizing Climate Finance from the Private Sector — Summary of GEF Experience. GEF/C.49/Inf.13.

Gupta S., J. Harnisch, D. C. Barua, L. Chingambo, P. Frankel, R. J. Garrido Vázquez, L. Gómez-Echeverri, E. Haites, Y. Huang, R. Kopp, B. Lefèvre, H. Machado-Filho and E. Massetti. 2014. *Cross-Cutting Investment and Finance Issues.* In Climate Change 2014: Mitigation of Climate Change. Contribution of Working Group III to the Fifth Assessment Report of the Inter-Governmental Panel on Climate Change, O. Edenhofer, R. Pichs-Madruga, Y. Sokona, E. Farahani, S. Kadner, K. Seyboth, A. Adler, I. Baum, S. Brunner, P. Eickemeier, B. Kriemann, J. Savolainen, S. Schlömer, C. von Stechow, T. Zwickel, and J. C. Minx (eds.), Cambridge, UK and New York, NY: Cambridge University Press.

International Energy Agency (IEA). 2015. *World Energy Outlook (WEO) Special Briefing.*

Micale, V., G. Frisari and F Mazza. 2013. Mapping the World Bank Group Risk: Mitigation Instruments for Climate Change. Climate Policy Institute (CPI). At: http://climatepolicyinitiative.org/wp-content/uploads/2013/09/World-Bank-Group-Risk-Mitigation-Instruments-for-Climate-Change-Brief1.pdf.

New Climate Economy (NCE). 2014. *Better Growth, Better Climate: The New Climate Economy Report.* At: http://newclimateeconomy.report/2014/

Organisation for Economic Co-operation and Development (OECD). 2015. Climate Finance in 2013–14 and the USD 100 billion goal. A report by the OECD in collaboration with Climate Policy Initiative (CPI).

Okereke, C. 2007. An Exploration of Motivations, Drivers and Barriers to Carbon Management: The UK FTSE 100. *European Management Journal* 25(6), 475–486, ISSN 0263-2373, http://dx.doi.org/10.1016/j.emj.2007.08.002.

UNEP DTU NAMA Pipeline Analysis and Database. 2015. At: http://www.namapipeline.org.

United Nations Framework Convention on Climate Change (UNFCCC). 2008. *Bali Action Plan.* Decision 1/CP.13. FCCC/CP/2007/6/Add.1. United Nations Framework Convention on Climate Change Secretariat.

United Nations Framework Convention on Climate Change (UNFCCC). 2015. At: http://unfccc.int/focus/mitigation/items/7172.php.

UNFCCC NAMA Registry. 2015. At: http://www4.unfccc.int/sites/nama/SitePages/Home.aspx.

Waissbein, O, Y., Glemarec, H. Bayraktar and T. S. Schmidt. 2013. Derisking Renewable Energy Investment: A Framework to Support Policymakers in Selecting Public Instruments to Promote Renewable Energy Investment in Developing Countries. United Nations Development Programme. New York, NY.

Chapter 5

Green Bonds and Climate Finance

Christa Clapp

Center for International Climate and Environmental Research — Oslo (CICERO), Gaustadalléen 21, 0349 Oslo, Norway
christa.clapp@cicero.oslo.no

Kamleshan Pillay

Center for International Climate and Environmental Research — Oslo (CICERO), Gaustadalléen 21, 0349 Oslo, Norway
kamleshan.pillay@cicero.oslo.no

Abstract

Green bonds apply environmental labelling to traditional bonds for financing green and climate projects. The green bond market has grown rapidly in recent years, and shows promise for furthering climate action. Multilateral development banks have been most active in issuing green bonds to support their environment and development agendas, but corporations and municipalities are also beginning to issue green bonds. To what extent can green bonds provide finance for climate action in developing countries? Emerging economies are becoming more active in the green bond market. The case of Johannesburg, the first city in an emerging country to issue a green bond, is examined for potential replication. Challenges and opportunities relating to green bond demonstration and governance, de-risking, and green integrity are also explored.

1. Introduction

While much of the investor news related to climate change revolves around divestment strategies and the carbon bubble, a quieter trend in labeled green investment tools such as green bonds is gaining traction. Green bonds are an example of a new green investment product that has enabled more proactive climate investment strategies, according to a report prepared for the German G7 Presidency [CICERO and CPI, 2015].

In the last two years, and particularly in the lead-up to the UN Climate Summit in New York in 2014, institutional investors have announced pledges for purchases of green bonds. Five major institutional investors have all announced targets of USD 1 billion green bond portfolios each [CBI, 2015a]. Further pledges have been made for low carbon portfolios, which could also include green bonds.

Several green bond indices also debuted in 2014 and 2015, including Barclays MSCI and Standard & Poor's (S&P). Oslo Securities Exchange created the first separate green bond listing in 2015. These aggregations allow for investors to more easily identify and target green investment products in their portfolios.

The green bond trend is growing in developing countries as well. Since the debut of the market in 2008, the World Bank has been active in developing new green bond issuance and reporting. Multilateral development banks now make up a significant portion of the green bond market. In addition, the Green Climate Fund (GCF) recently announced that some of the initial proposals to be funded include green bonds for energy efficiency projects in Latin America.

Emerging economies are beginning to be active in the green bond market, including Brazil, China, India, Korea, Mexico and South Africa. China's government has taken a particular focus on green investment, with the intent to establish a green bond market. China's central bank, the People's Bank of China, has developed green definitions. In India, the private bank Yes and the Export Import Bank of India have issued green bonds. In 2014, the city of Johannesburg in South Africa issued the first municipal green bond in an emerging market. Nafin, the state-owned development bank of Mexico, the

Brazilian food producer BRF, and the Export-Import Bank of Korea (KEXIM) have also issued green bonds within the last three years.

This chapter explores many of the questions surrounding the rapid green bond market growth, including: what is a green bond, and how do we know if it is green? The potential opportunities and challenges for channelling financial flows through green bonds to support climate change action in developing countries is examined, with a focus on the city of Johannesburg green bond as a case study.

2. Green Bonds: Description and Definitions

Green bonds are a relatively new way of packaging a traditional financial instrument directed towards activities with a positive environmental outcome. The global green bond market has grown rapidly in response to investor appetite. There are several initiatives that support environmental robustness and transparency, but green bonds face ongoing questions about environmental integrity.

2.1. *What are green bonds?*

In the financial sector, a bond is categorized as a debt instrument, which allows the issuer to raise capital from investors. Bonds can be issued by either public or private institutions. The fixed returns of a bond structure can be attractive to institutional investors, which control the bulk of global investment financing.

Green bonds are simply bonds that finance green or climate change activities. They can be issued by a range of issuer types, such as multilateral development banks, municipalities, and companies seeking to raise their environmental reputation. Green bonds provide an opportunity to provide finance to an aggregation of smaller projects. This aggregation consolidates risk compared to direct loans that have full exposure to the risk of a particular project [Mathews and Kidney, 2014]. Part of the appeal of green bonds to investors lies in the fact that they are a common well-known financial instrument, labelled to indicate their focus on environmental issues. Purchasers of green bonds include institutional investors from the public and private pension funds and insurance companies. A few green bonds have also been issued for individual investor purchase.

The term "green bond" is typically used to indicate a bond that supports climate change or other environmental projects. Most of the green bonds issued to date have focused on climate change projects. Some green bonds also include consideration of other environmental issues, such as water conservation or biodiversity.

The proceeds of green bonds have been used to finance both mitigation and adaptation projects, but with the majority focusing on mitigation. The multilateral development bank green bonds have financed mainly renewable energy projects [BNEF, 2014]. Some municipalities have included both mitigation and adaptation components in their green bonds. A typical focus of adaptation projects is water management. For adaptation or climate-resiliency projects in developing countries that are less profitable, other financial sources are likely to be more appropriate [Kato *et al.*, 2014].

Green bonds can be used to finance climate activities in developed and developing countries alike. Multilateral development banks have been very active in issuing green bonds to finance climate projects in developing countries. Corporate and municipal issuers have been more active in developed countries; however there could be opportunities for further issuance in developing countries.

2.2. *State of the green bond market today*

The World Bank and the European Investment Bank were early issuers of green bonds to target climate activities in developing countries, driven by demand for sustainable investment products from public pension fund investors in Europe. In 2008, the World Bank issued the first prototype green bond, working together with Skandinaviska Enskilda Banken (SEB) to develop its framework. The Center for International Climate and Environmental Research — Oslo (CICERO)[1] was asked to provide an independent evaluation, or second opinion, on the environmental integrity of the green bond

[1] A non-profit climate research center based in Oslo, Norway (authors' employer).

framework. The World Bank continues to be a market innovator, leading the way on reporting for green bonds (see Box 1 for further information). Around the same time as the initial World Bank green bond, the European Investment Bank issued its first Climate Awareness Bond, and has also continued to be active in the green bond market.

In the first few years, multilateral development banks were the drivers of the market. The term "green bond" became a common term to indicate bonds linked to both climate and environmental activities. By 2011, corporations and municipalities saw the potential value of attracting new investors by labelling green bonds.

The green bond market has experienced rapid growth in the last two years (see Figure 1). The green bond market was valued in mid-2015 at USD 66 billion, with new issuance having tripled in 2014 over 2013 [CBI, 2015a]. Multilateral development banks are still a key driver of the market, but corporations and municipalities have taken the market to a new level.

Yet to put these billions in context, the green portion of the market is less than one-tenth of a percent of the total global bond market, which is estimated at approximately USD 100 trillion [Kato *et al.*, 2014; OECD, 2015]. This points to a large upside potential if, for example, green bonds are used to finance green infrastructure projects in a widespread manner. A recent study by CICERO extrapolates that if the labelled green bond market were to grow by an order of magnitude to USD 1 trillion in 2020, global CO2 emissions could be reduced by 100 MT-CO2 from business-as-usual projections in 2030 [Glomsrød and Taoyuan, 2015].

Governments can play an active role in supporting the financing of green projects through green bonds. Approximately 75% of green bonds and identified climate-themed bonds to date have been issued by government owned or backed entities, primarily though developed country governments, such as Canada, France, US and UK, and multilateral development banks [CBI, 2014a]. In addition to green bond issuance, governments could also play a role in risk mitigation for green bonds, i.e., to reduce risks related to e.g., currency fluctuation,

Box 1: World Bank Green Bonds

Since the first issuance in 2008, the World Bank has mobilized financing of over USD 7 billion through green bond issuance in 18 different currencies, diversifying and expanding their investor base [World Bank, 2015a]. Given the institution's focus on poverty reduction, all of the financed projects have a development component, but with clear climate change mitigation and/or adaptation benefits. Projects are financed in the following sectors:

- Mitigation
  - Renewable energy
  - Energy efficiency
  - Sustainable transport
  - Waste and waste water
  - Agriculture and land use
  - Forestry
- Adaptation
  - Waste and wastewater systems
  - Infrastructure, energy and built environment
  - Agriculture and ecological resources

The World Bank is also a market leader in impact reporting on green bonds, listing some expected impacts from projects on their Treasury website

(http://treasury.worldbank.org/cmd/htm/MoreGreenProjects.html)

and highlighting selected project impacts in their annual green bond newsletter. The reported impacts include greenhouse gas (GHC) reductions but also other indicators such as land area covered by flood protection [World Bank, 2014]. The World Bank is also coordinating with other development finance institutions on impact reporting.

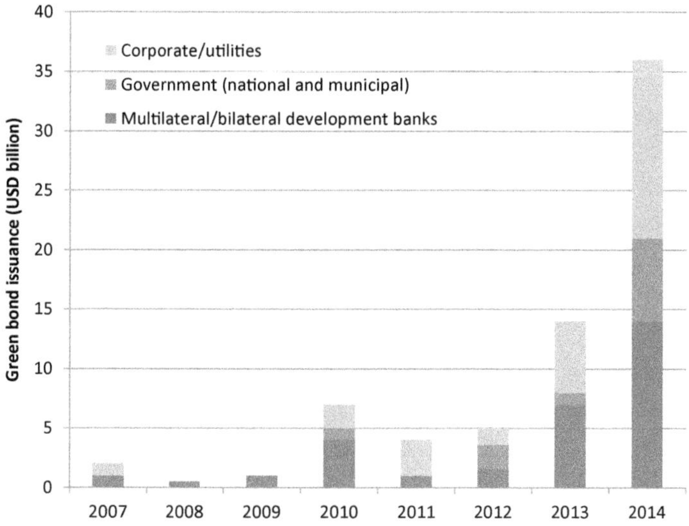

Figure 1. Green Bonds by Issuer Type.
Sources: BNEF [2014], CBI [2014a, 2015a] and World Bank [2015a].

political risk, and credit risk by providing insurance on green bonds [CICERO and CPI, 2015].

Multilateral and bilateral development finance institutions have issued 43% of the green bonds from 2007 to 2014, raising approximately USD 30 billion in financing for climate change projects in developing countries. Alongside many multilateral development banks, the African Development Bank (AfDB) initiated a green bond program in 2013 (see Box 2). Green bonds have also been issued by government export credit agencies, such as Export Development Canada and the KEXIM.

In 2013, local governments began to issue green bonds for climate-related activities or lending. At the local level, government green bond issuance has primarily been in developed countries, e.g., in Sweden and Canada. A notable exception is the city of Johannesburg, South Africa, which was the first C40 city to issue a green bond, valued at USD 138 million [National Geographic, 2014]. The Johannesburg bond finances a biogas-to-energy project, a solar heating initiative, and dual-fuel buses. The Johannesburg bond illustrates

Box 2: African Development Bank Green Bond Program

One of the primary objectives of the African Development Bank (AfDB) is to create inclusive growth that is sustainable and ensure that Africa transitions to a green economy where the food and water security, energy and natural resources, job creation and economic development are all protected. The AfDB Green Bond is one of the programs implemented by the bank that aims to facilitate green economic growth through the financing of green and climate change projects. Green bonds from the AfDB are Aaa/AAA/AAA rated by all the major credit rating agencies and carry a 0% risk weighting under the Basel II and Basel III recommendations for banking regulations.
Eligible projects include:

- *Energy*: renewable energy generation, energy efficiency improvements in lighting and electricity transmission
- *Transportation*: vehicle efficiency and urban transport
- *Waste*: solid waste management
- *Industry*: process efficiency improvements, fugitive emission capture
- *Forestry*: reduced deforestation
- *Water*: drainage systems for flood protection
- *Agriculture*: water supply and access

a direct action from a developing country that attracts private investments (see Section 3.2.3 for the city of Johannesburg case study).

Corporations have played an increasing role in green bond issuance since 2011. Transportation and real estate companies, as well as corporate banks, have issued green bonds. Most of these corporations have been based in developed countries, although some green bonds provide financing for activities based in developing countries. Yes Bank in India also issued a green bond in 2015.

On the demand side, investor interest has been increasing. Investing in green bonds can be part of a pro-active strategy to mitigate climate risk to investments, and green bonds can easily be incorporated into existing investment portfolios. While divestment campaigns focus on the exclusion of specific companies or industries from investment portfolios, an inclusive strategy to factor in risks from physical climate change impacts and changing carbon policies can include the purchase of green bonds [Clapp *et al.*, 2015].

Many green bonds issued in the past few years have been over-subscribed, i.e., investor demand was greater than the amount of the bond issued. The World Bank, KfW, and Yes Bank of India all increased the size of the bonds they issued in response to investor over-subscription [CBI, 2015a].

Bonds generally have a lower risk profile than equity instruments due to their fixed returns. This makes bonds particularly attractive for institutional investors to provide steady payouts over a long-term investment horizon. Organisation for Economic Co-operation and Development (OECD)-based pension funds invested an average of 53% of their portfolios in bonds; OECD-based insurance companies invested 64% in bonds [OECD, 2015].

Institutional investors have been increasingly vocal regarding their demand for low carbon investment products, including green bonds. Following the United Nations (UN) Climate Summit in 2014, institutional investors (including ACTIAM, Barclays Bank, Deutsche Bank, KfW and Zurich Insurance Group) pledged their intent for combined investment of over USD 5 billion in green bonds [CBI, 2015a].

2.3. *Many shades of green*

There is no agreed definition or standard or what types of projects can be financed by a green bond. On the one hand, the lack of common definitions gives more flexibility in bringing green bonds to the market, with no strict requirements. However, concerns about varying degrees of transparency and the potential for "green-washing" need to be addressed, especially as more corporations issue green bonds. Climate science can help guide investment choices, including for green bonds, which should avoid locking-in infrastructure that is not supportive of a low carbon climate-resilient future [Clapp *et al.*, 2015]. Environmental safeguards need to protect against potentially competing interests, and there are several trends in the market that support some level of environmental integrity.

There are both labelled and unlabelled green bonds in the market. Labelled green bonds are clearly identified as green, but with varying definitions of project types and degrees of independent review on environmental quality. There are also bonds that finance climate action but are not labelled specifically as green. The broader universe of both labelled and unlabelled bonds that has been identified as climate-themed is estimated in value at approximately USD 600 billion [CBI, 2015a].

A group of underwriting banks came together in 2014 to sign the Green Bond Principles, which was updated in 2015 [ICMA, 2015]. This document outlines voluntary guidance for issuing green bonds, focusing on delineated accounting and transparency. Yet the Principles say little on what green project types should be eligible.

Shareholders are also paying more attention to the environmental integrity of green bonds. Over 25 investors, including some of the largest institutional investors, signed a Statement of Investor Expectations for the Green Bond Market [Ceres, 2015]. The Statement pushes further than the Green Bond Principles, noting that the signatory investors will exercise additional diligence when evaluating bonds that finance environmentally marginal projects. The signatory investors also expect annual impact reporting and note the need for independent assurance or auditing on the selection and tracking of green projects. At the Paris Conference of Parties (COP), 27 investors (representing over USD 11 trillion) signed the Paris Green Bonds Statement, committing to growing the market and calling for greater transparency and clear standards [CBI, 2015b].

While some limited progress has been made in developing a global standard for green bonds, the application of a one-size–fits-all standard may be difficult in practice. One initiative to develop a global standard has been instigated by a UK-based charity, the Climate Bonds Initiative. Four bonds have been certified according to the Climate Bonds Standard to date.

A range of best practices or standards may emerge in different regions in the coming years. China is the first country to develop a national green bond standard, showing an active government role in defining what is green. The Green Projects Catalogue, developed in December 2015 by the People's Bank of China, provides a list of standards by sector for screening green bonds.

In the absence of more specific guidance on what is a green project outside the Chinese market, the practice of using of independent reviews, or "second opinions", has become more common. Second opinions are offered by an independent party on the issuer's green bond frameworks at the time of issuance. There is a wide range of second opinion providers active in the market, with varying approaches and details in what they review. Some focus on potential climate and environmental impacts, others encompass social governance issues as well, while still others focus on cash flow accounting of the bond proceeds.

The Green Bond Principles recommend the use of external assurance, such as second opinions, audits or certifications. Further, the Oslo securities exchange, Oslo Børs, launched a separate green bond listing in 2015, requiring a public second opinion for an issuance to be listed.

Despite these calls for external assurance, there are still issuers that self-label what constitutes a green bond, forgoing the use of an independent second opinion. Approximately 40% of the green bond market has not undergone an independent review [CBI, 2015a].

Several green bond indices have also recently emerged, allowing investors to quickly track green bond investments. Yet each index follows different logic for defining which green bonds can be included. For example, the Barclays MSCI green bond index automatically excludes green bonds that finance hydro projects larger than 25 MW, in response to concerns of local environmental and social impacts, regardless of independent second opinions on expected impacts from planned large hydro projects.

In terms of reporting, the green bond market is flexible, with no strict requirements. The Statement of Investor Expectations calls for impact reporting. The Principles also recommend the use of qualitative, and where possible, quantitative reporting (e.g., on greenhouse gas emission reductions), and welcome initiatives to harmonize reporting. The current market practice is for the issuer to annually report a sample list of projects financed by the green bond. Impact reporting for green bonds is still in an early phase, with the World Bank providing leadership and working with other issuers on harmonization of reporting (see Box 1).

3. Climate Finance for Developing Countries

Over 40% of the green bond market has been issued by multilateral and bilateral development finance institutions, supporting climate action in developing countries. There have also been several issuances by government-backed entities in emerging economies, including national development banks (NDBs) and municipalities.

The case of the city of Johannesburg's green bond issuance provides insights as to the motivations and design of a municipal green bond. Yet the African bond market remains underdeveloped, particularly at the municipal level. Although there are opportunities to scale-up green bond issuance to some degree in developing countries, barriers remain in relation to lack of knowledge of the green bond market and supporting policies and regulations. Possibilities for governments to support green bond development in developing countries include demonstration and governance, de-risking, and supporting green integrity.

3.1. *Are green bonds supporting action in developing countries?*

In the context of the international climate negotiations, climate finance typically refers to the pledge of USD 100 billion from developed countries to support climate action in developing countries by 2020. Yet there are no universally agreed definitions for what constitutes a financial transfer that could 'count' towards the USD 100 billion. Furthermore, it is complex to understand the origins of the buyers of green bonds, e.g. if they are government owned pension funds or private investors and from which region. The data on investors of green bonds is patchy and often not publically available.

In general terms, however, it is useful to consider if green bonds are supporting climate action in developing countries, where financing is more sorely needed. The green bonds issued by multilateral and bilateral development finance institutions can serve as a first-order approximation of green bonds supporting action in developing countries. These institutions issued 43% of the green bonds since 2007, for a total amount of USD 30 billion. However only a fraction of bondholders' information is publically available for multilateral development banks, since many investors are located outside of the US and EU where disclosure is required [BNEF, 2014]. Data is available for the Kexim bond (issued in US dollars), however, showing that 79% of the investors were located either in the US or Europe, indicating a more transparent flow of financing from developed countries to an emerging market [Wee, 2013].

In addition to the multilateral and bilateral development bank issuance, a few green bonds have been issued by government agencies, municipalities and corporations based in emerging countries, including e.g. the city of Johannesburg in South Africa, Yes Bank in India, and Nafin, the state-owned development bank of Mexico (see Table 1). However, the data on investors for these green bonds has not been made publically available, so it is not possible to fully trace the flow of investment from region of origin. However, we do know that some of the bonds have been issued in Euros or dollars to attract overseas investors, e.g. food producer in Brazil BRF and export-import bank of Korea Kexim.

Many international corporations also include operations in developing countries, some of which are included in green bond frameworks. For example, GDF Suez issued a green bond that supported a hydro dam in Brazil (which was met with some controversy over local environmental and social issues — for further discussion of green integrity, see Section 2.3 on "Many Shades of Green"). However reporting on what portion of a green bond supports projects in developing countries varies widely with each issuer.

Table 1. Green Bond Activities in Emerging Countries (beyond Multilateral Development Banks).

Green bond issuance and activity (beyond multilateral and bilateral development banks)		
Country	**National and local government-backed institutions**	**Private sector**
Brazil		• BRF food producer issuance • FEBRABAN Brazilian Association of Banks created Green Bond Market Development Committee
China	• People's Bank of China sectoral standards	
India	• Export Bank of India issuance • Industrial Development Bank of India (IDBI) issuance	• Yes Bank issuance
Korea	• Kexim export-import bank of Korea issuance	
Mexico	• Nafin development bank issuance	
South Africa	• Industrial Development Corporation issuance in cooperation with Nedbank • City of Johannesburg issuance	

Source: CBI [2015a].

Another issue of debate in the international climate negotiations is whether climate finance is "new", i.e., is additional to previous financial flows for climate activities. It is very difficult to determine if green bonds raise "new" financing in relation to what financing may have been available in the traditional bond market. This is in large part due to a lack of comprehensive and detailed data on financial transactions and their attribution to climate action.

However, there is anecdotal evidence that some new investors have been attracted to green bond issuance by the World Bank and the International Finance Corporation (IFC) [World Bank, 2015a]. The diversified and expanded investor base includes institutional investors with specific sustainability or responsible investment objectives that have not previously purchased bonds from these institutions.

Yet it is unclear if this indicates new financing for climate action or re-packaging of existing financial products [CICERO and CPI, 2015]. Many bonds and green bonds are used to re-finance existing projects, providing an important vehicle for climate financing, but perhaps not indicating additional finance. Re-financing for green projects can help increase investor confidence in the market and thereby lead to additional green projects [CBI, 2014a].

3.2. *Case study: Johannesburg green bond and the African context*

3.2.1. *Context*

According to the World Bank, infrastructure investment will be vital if Africa is to meet the continent's development targets. Furthermore, estimates suggest that all African countries would need to realize a 2.2% increase in economic growth (per capita) if they are to improve their existing infrastructure. The actual cost of meeting Africa's infrastructure shortfall is estimated to be USD 38 billion dollars per annum with a further USD 37 billion dollars per annum required for operation and maintenance of existing infrastructure [World Bank, 2015b]. Besides the issue of infrastructure investment, climate change is another concern that the continent faces. According to the 5th

Assessment Report of Intergovernmental Panel on Climate Change (IPCC), Africa is suggested to be the most vulnerable continent to the effects of climate change. This is owing to a lack of financial capacity, technical expertise and infrastructure. Furthermore, climate adaptation costs are expected to rise to USD 50 billion dollars per year by 2050 [UNEP, 2015]. Considering these factors, African countries will need to invest in a low carbon development pathway with climate resilient infrastructure.

With reference to the South African context specifically, the country uses a large amount of energy, is highly energy-intensive, and heavily dominated by the extraction of raw materials and primary processing [Nkomo, 2005]. In recent times, the country has suffered severely from load shedding owing to aging infrastructure and demand not meeting supply of electricity. The country is also considered to be the 12th largest emitter of CO_2 emissions in the world and the largest on the African continent. The country is responsible for nearly half the CO_2 emissions for the entire continent of Africa, and about 1.6% of global emissions [IETA/CDC Climate Research/EDF, 2014]. The Government of South Africa has also stated that over the next three years from 2013 to 2014, ZAR 827 billion (approximately USD 57 billion) will be invested in building new and upgrading existing infrastructure. Despite this significant investment over the next three years, it is uncertain as to whether these large investments into infrastructure will continue into the future (particularly considering the expectation that modest growth is expected for the South African economy). Consequently, it is likely that South Africa will need to look towards new sources of finance as well as the use of different financial instruments to stimulate capital.

3.2.2. *Bond markets in Africa*

In many African countries, loans are the preferred instrument to raise capital rather than issuing debt in the form of bonds [Oji, 2015]. Yet bonds have a distinctive advantage over loans as they are long-term securities and consequently suffer less from short-term volatility. Furthermore, municipal and government bonds are considered to be more secure with higher credit ratings than other types of issued debt investments such as corporate bonds, due to a significantly lower likelihood of default. Bonds are also highly liquid on secondary markets even though these markets are quite under-developed in Africa (and also under-developed for the green bond market). Entities which have purchased bonds are also protected by law and unforeseen circumstances such as bankruptcy of the issuer. Lastly, bonds issued at a national or city level may possess tax exemptions and lower interest rates and longer repayment periods (interest on municipal bonds is tax deductible in South Africa). This is owing to the fact that bond parameters are determined by the issuer rather than loan parameters that are determined by the bank.

In Africa, financial markets are under-developed, a lacking requirement if bond transactions are to occur at a larger scale. However, it is possible for African governments and cities to take steps to support the growth of the bond market. Possible steps include the communication of issuance programs, the creation of a platform to support issuances (and create benchmarks), the creation of sustainable fiscal policy and the removal of regulatory costs for corporate issuances. In South Africa, municipal bonds with maturities of more than a year are issued by cities for the primary reason of raising capital for development projects [SACN, 2008]. It must be noted that South African city bonds are not guaranteed by the South African national government. Other risks that may hinder the issuance of green bonds in Africa, as well as other regions, include the various types of risk (inflationary, credit, liquidity and interest rate risk) [African Progress Panel (APP), 2015].

There has been little activity in the municipal bond market in Africa. Since 2004, the City of Johannesburg (COJ), South Africa, has issued seven bonds with the aim of stimulating capital of ZAR 8 billion. Six of the seven of bonds issued by the COJ are unsecured, however they are tax deductible. The City of Tshwane (Pretoria) has managed to raise ZAR 1.39 billion by issuing two unsecured bonds. South African local governments are not the only issuers of bonds in Africa, with Nigeria and, to a lesser extent, Kenya, also possessing a local government bond market [Oji, 2015]. In Nigeria, from the period of 1986 to 2012, 20 states have issued bonds to finance development projects. Nigeria's local and state issued bonds are tax deductible however they are not guaranteed by the Nigerian government. In reality, they are backed by taxes and income from completed projects such as road tolls. Even though Diaspora, Islamic and sovereign bonds have been issued in other parts of Africa (including Ethiopia, Gambia, Sudan and Ghana), city bonds or local government bond issues are scarce.

3.2.3. *The City of Johannesburg green bond issue*

The City of Johannesburg (COJ) has set out a long-term strategy that supports the transition to a low carbon economy, outlined in the Integrated Development Plan (IDP) of the city. Stipulated within the IDP, the city council aims to:

- Define its role in the green economy, while supporting creative economic sectors;
- Build new infrastructure considering energy efficiency; and
- Make considerable investment towards green infrastructure.

At the C40 Cities Mayors' Summit held in February 2014, Mayor Parks Tau of the (COJ) stated that the issue of a green bond would help facilitate the access to finance for smaller and medium enterprises that want to retrofit or develop new buildings and technologies in relation to climate change. Previously, approximately 87% of electricity in Johannesburg was generated from coal. In recent years, the city council has managed to reduce this reliance on coal by 20% through appropriate demand-side management (DSM) implementation and increased energy efficiency [Dlamini, 2014]. The stimulation of investment from the green bond will further help to reduce the reliance on coal.

The bond was issued in 2014, valued at ZAR 1.46 billion (see Table 2). The bond was 1.5 times oversubscribed, reflecting a high level of investor interest. The proceeds from the green bond will be used for financing green initiatives and climate change mitigation

Table 2. City of Johannesburg Green Bond Statistics.

Green City Bond Stats [COJ, 2014]:		
Issue Date	**Yield**	**Use of Proceeds**
June 2014	1.85%	
Size	**Rating**	Renewable Energy,
1.46 billion ZAR	BBB (Fitch)	Other Energy
Tenor	**Underwriter**	and Transport
10 years	Standard Bank Group	

Sources: COJ [2014].

strategies that will reduce greenhouse emissions [Brand, 2014; SA Building Review, 2015; Williams, 2014]. Projects eligible for financing include:

- *Renewable energy*: solar-panel power generation, water heaters, and biogas-to-energy projects;
- *Other energy*: generation of electricity from modified waste-water plants, landfill methane gas capture and energy generation, and other biogas-to-energy projects; and
- *Transport*: funding for the purchase of 152 hybrid-fuel buses [CBI, 2014c; SA Building Review, 2015; Williams, 2014].

Several other green bonds have also been issued in South Africa. In November 2015, the IFC, which is a unit of the World Bank, issued the first rand-denominated green bond. The bond will raise approximately USD 71 million in climate friendly investments [Reuters, 2015]. The tenor of the bond is nine years with a yield of 8.72%. Similar bonds have been issued by the IFC in Chinese renminbi and Indian rupees. There have been two other cases of bond issuances in South Africa for green investment. In 2012, the Industrial Development Corporation of South Africa (IDC), in collaboration with Nedbank, launched a ZAR 5 billion bond, which raised capital to fund infrastructure for renewable energy [UJUH, 2012]. In 2013, Soitec issued a ZAR 1 billion bond to fund the development of a 44 MW PV solar power plant [GBG, 2013]. Although it was not expressly labeled as a green bond, the Soitec bond was the first bond to finance a solar power plant based on concentrator photovoltaic (CPV) technology.

3.3. *Challenges and opportunities*

South Africa's bond market is considered to be the most mature within Africa; therefore it is difficult to draw broad regional conclusions from the COJ's green bond issuance. Yet overall there is appreciable potential for increased green bond issuance within South Africa. Despite maturity of the bond market in South Africa, enhanced awareness amongst market players is needed to create a more active domestic green bond market. Moreover, more effort needs to be placed on explaining and understanding the green bond concept and the benefit of their issuances. Lastly, concerns about meeting the green bond reporting guidelines may also inhibit the issuance of bonds [Mda, 2015].

More broadly, developing countries face several major barriers to issuances of green bonds, including:

- A lack of policy and regulatory frameworks within secondary markets;
- A lack of awareness about the benefits of bonds;
- An absence of understanding on the processes needed to reform, structure and develop a market to stimulate bond issuances; and
- The issue of credit rating: If green bond issues were to increase at a municipal level, then measures need to be implemented to maintain a suitable credit rating [SACN, 2008].

To address these challenges, governments and other actors have several possible ways to engage. Some of these opportunities for supporting green bonds are described below, including: green bond demonstration and governance; de-risking and credit enhancement; and green integrity and transparency. The opportunities are described here with a view to developing countries; however similar ideas could be useful to the global green bond market as well.

3.3.1. *Demonstration and governance*

Investors are indicating increasing demand for green bonds, putting pressure on the project pipeline for green financing. One potential challenge is to identify projects and activities that could be eligible for green bond financing, as well as identifying existing projects that could be re-financed through green bonds. While many projects should be available in theory, projects with predictably steady returns are more conducive to bond financing.

Another potential challenge is the aggregation of small-scale actors and projects. For example, the dominance of smallholder farmers in Malawi may result in higher transaction costs for a green bond issuance, and might be better served through microfinance. Alternately, smallholder entities could aggregate through an organization or commercial bank, allowing for a green bond to be issued at the appropriate scale.

Government entities and municipalities can issue green bonds in developing countries, but may face limitations in the governance and capacity to organize projects and the reporting for green financing. Countries with more widespread governance problems, e.g., in less developed countries, will likely face greater challenges in issuing green bonds from government institutions. However, some emerging economies are taking the initiative to organize for attracting green bond financing. For example, the Brazilian Association of Banks has established a Green Bond Market Development Committee [CBI, 2015a].

Corporations can also take the initiative to demonstrate green finance in emerging countries. Yes Bank in India and Brazilian food producer BRF are two examples where corporations issued green bonds to overseas investors. The BRF bond was issued in euros to attract European investors.

3.3.2. *De-risking investments*

Development finance institutions can continue to play a leading role in identifying projects and de-risking investments. Green bonds issued by institutions such as the ADB and the World Bank help demonstrate the viability of green bond financing in developing regions (see Boxes 1 and 2). The high credit ratings of multilateral finance institutions attract large institutional investors, and can also be used to mitigate the risk of green bonds issued by entities in developing countries. For example, the IFC issued a green bond listed on the London Stock Exchange to finance its purchase of shares of Yes Bank's green bond. The Inter-American Development Bank is establishing a regional Energy Efficiency Green Bond Facility, as one of the initial eight projects to receive funding from the GCF.

NDBs also have a critical role to play in strengthening the green bond market, as they are able to catalyze and complement private sector investment. Approximately 16% of all NDBs worldwide hold assets equivalent to USD 10 billion or more [De Luna-Martinez, 2012]. Their status within local markets and local private sector and their ability to interact with government agencies make them the ideal candidate to act as intermediaries to scale up climate finance initiatives [Smallridge *et al.*, 2012]. NDBs have a development approach which allows them to identify priority sectors that green bonds could be directed towards. Lastly, NDBs could have a leveraging effect mobilizing capital from commercial banks. An example of a NDB issuing a green bond was the KfW's inaugural issue valued at EUR 1.5 billion in July 2014 (0.375%, AAA rating) [CBI, 2014b]. The vision of the bank, which is further reaching than individual capital market participants, could result in a greater effect on climate change protection. The KfW bond was oversubscribed with the bank receiving EUR 2.65 billion in orders.

Investors face a higher risk when buying bonds from developing countries, which is one of the barriers to green bond implementation. Consequently, it may be useful to investigate financial instruments that can complement green bond issuances such that they are more attractive to investors. Additional instruments could be used to improve the credit worthiness and thereby support the issuance of green bonds. Credit enhancement refers to the reassurance that the borrower (green bond issuer) will commit to the repayment of the bond through additional collateral, insurance, or a third party guarantee.

3.3.3. *Green integrity*

For all green bonds, regardless of geographical location, there exists a potential risk for "greenwashing", or labeling the bond green when it supports projects that have negative impacts on the environment or climate. While this is always a risk, it may be a greater risk as corporations issue more green bonds, since they may not have as rigorous environmental standards and reporting as some government agencies or multilateral finance institutions. The credibility of the green bond market hinges on financing for robust green projects.

Independent evaluations and standards, as well as a high level of transparency can support environmental credibility for green bonds issued by all institutions and corporations. One approach to prevent greenwashing is to take a top-down approach with government developing standards and procedures. China is the first country to take such an approach, with green bond standards recently developed by the People's Bank of China.

Regular reporting efforts enhance transparency of green bonds. The World Bank is working with other green bond issuers to support and establish reporting processes. Further demonstration and capacity building on reporting could support green bond issuance in developing countries.

4. Concluding Remarks

The green bond market is in a period of rapid growth. As more investors take environmental factors into consideration, demand is increasing. The types of issuers coming to market represent a broader range of public and private actors, and across a broader geographical range, than even just a few years ago. Green bonds have primarily focused on mitigation projects to date, but some adaptation projects are beginning to be included in green bond frameworks.

Green labeling and transparency are important for the integrity of the market, especially as the market includes new issuance from corporations that are just beginning their practices to track environmental impacts. While it is difficult to imagine one global standard, a series of best practices may emerge in different regions. For the credibility of green bonds, these practices should be linked to independent evaluations, standards and reviews.

Some of the green bond issuers (e.g., IFC, World Bank) have noted that new investors, focused on environmental and sustainable investing, have been attracted to the green labeling. Definitions and data are too patchy to determine if this constitutes "new" climate finance. However, green bonds that re-finance green projects also play a valuable catalytic role in liquidity for the green bond market.

A significant portion of green bonds has supported climate action in developing countries, primarily through multilateral and bilateral development finance institutions. Inaugural corporate and municipal bonds have also been issued in India and South Africa.

The case of the city of Johannesburg green bond shows initiative to seek green financing for efficient retrofitting, but municipal bonds face an immature market in Africa. Challenges to green bond issuance in Africa and other developing regions include underdeveloped bond markets, lack of awareness of green bonds, and a lack of supporting policies and regulations.

There is scope for developing countries to take advantage of green bond financing in several ways. International financial institutions such as the ADB and the World Bank can continue to scale up their green bond activities. National and local governments can raise awareness of green bonds through project demonstration and governance, mitigate risk through credit-enhancement, and support efforts to improve the environmental integrity and transparency of green bonds.

References

African Progress Panel (APP). 2015. *Power People Plant: Seizing Africa's Energy and Climate Opportunities.* Research Report, 182.

Bloombery New Energy Finance (BNEF). 2014. *Green Bonds Market Outlook 2014*. June. At: http://about.bnef.com/white-papers/green-bonds-market-outlook-2014/.

Brand, R. 2014. *Johannesburg Plans Green Bonds*. Moneyweb [Internet]; [updated 2015 August 21]. At: http://www.moneyweb.co.za/archive/johannesburg-plans-green-bonds/.

Center for International Climate and Environmental Research — Oslo (CICERO) and Climate Policy Initiative (CPI). 2015. Background Report for G7 on Long-Term Climate Finance. Prepared for the German G7 Presidency by CICERO and CPI. June. At: http://climatepolicyinitiative.org/publication/background-report-for-g7-on-long-term-climate-finance/.

Ceres. 2015. *A Statement of Investor Expectations for the Green Bond Market*. Ceres Investor Network on Climate Risk [Internet]; [updated 2015 February 10]. At: http://www.ceres.org/files/investor-files/statement-of-investor-expectations-for-green-bonds/at_download/file.

City of Johannesburg (COJ). 2014. *The City of Johannesburg Issues First Ever Green Bond*. COJ [Internet]; [updated 2015 August 26] http://www.joburg.org.za/index.php?option=com_content&view=article&id=9073:04062014-the-city-of-johannesburg-issues-the-first-ever-jse-listed-green-bond-&catid=217:press-releases-2013&Itemid=114.

Clapp, C., K. H. Alfsen, A. Torvanger and H. Francke Lund. 2015. Influence of Climate Science on Financial Decisions. *Nature Climate Change* 5.

Climate Bonds Initiative (CBI). 2014a. *Bonds and Climate Change: The State of the Market in 2014*. HSBC and CBI. At: http://www.climatebonds.net/bonds-climate-change-2014.

Climate Bonds Initiative (CBI). 2014b. *Germany's KfW Dives in Deep*. CBI. https://www.climatebonds.net/2014/07/germany%E2%80%99s-kfw-dives-deep-eur-15bn-2bn-5yr-aaa-green-bonds-go-germany.

Climate Bonds Initiative (CBI). 2014c. *First Emerging Market Green City Bond*. CBI. At: https://www.climatebonds.net/2014/06/just-out-first-emerging-market-green-city-bond-city-johannesburg-green-bond-approx-v15bn

Climate Bonds Initiative (CBI). 2015a. *Bonds and Climate Change: The State of the Market in 2015*. HSBC and CBI. At: https://www.climatebonds.net/resources/publications/bonds-climate-change-2015.

Climate Bonds Initiative (CBI). 2015b. *The Paris Green Bonds Statement, Climate Bonds Initiative*. At: http://www.climatebonds.net/files/files/Paris_Investor_Statement_9Dec15.pdf.

De Luna-Martinez, J. and C. L. Vicente. 2012. *Global Survey of Development Banks*. World Bank Working Paper 5969, 5969.

Dlamini, P. 2014. *Jozi Pioneers Green Bonds*. Times Live [Internet]; [updated 2015 August 21]. At: http://www.timeslive.co.za/thetimes/2014/02/06/jozi-pioneers-green-bonds.

Glomsrød, S. and W. Taoyuan. 2015. *Business as Unusual: The Implications of Fossil Divestment for Financial Flows, Economic Growth and the Energy Market*. Center for International Climate and Environmental Research —

Oslo (CICERO) and HSBC.

Green Business Guide (GBG). 2013. *The City of Johannesburg Issues the First Ever JSE Listed Green Bond.* GBG [Internet]; [updated 2015 August 5]. At: http://www.greenbusinessguide.co.za/standard-bank-group-arranges-first-renewable-energy-bond-in-south-africa/.

International Capital Market Association (ICMA). 2015. Green Bond Principles, 2015. At: http://www.icmagroup.org/Regulatory-Policy-and-Market-Practice/green-bonds/green-bond-principles/.

International Emissions Trading Associations (IETA), CDC Climat Research, and Environmental Defense Fund (EDF). 2014. South Africa: An Emissions Trading Case Study. IETA, CDC Climat Research and EDF, [updated 2015 May]. At: http://www/google.no/url?sa=t&rct=j&q=&esrc=s&source=web&cd=1&cad=rja&uact=8&ved=0ahUKEwijhcX2i83OAhVmYJoKHW1jA7IQFggiMAA&url=http%3A%2F%2Fwww.ieta.org%2Fresources%2FResources%2FCase_Studies_Worlds_Carbon_Markets%2Fsouth_africa_case_study_may2015.pdf&usg=AFQjCNGRuoep1rOVYMpJvzDsQ9j8R7zPNw

Kato, T., J. Ellis and C. Clapp. 2014. *The Role of the 2015 Agreement in Mobilising Climate Finance.* Organisation for Economic Co-operation and Development (OECD), November. At: http://www.oecd.org/env/cc/(2014)7%20The%20role%20of%20the%202015%20agreement.pdf.

Mathews, J. A. and S. Kidney. 2012. Financing Climate Friendly Energy Development through Bonds. *Development Southern Africa* 29, 337–349.

Mda, T. 2015. Personal Communication. Pretoria, South Africa: Treasury of South Africa.

National Geographic. 2014. *Johannesburg the First C40 City to Issue Green City Bond.* City Solutions, National Geographic [Internet], [updated 2014 June 12]. At: http://voices.nationalgeographic.com/2014/06/12/johannesburg-the-first-c40-city-to-issue-green-city-bond/.

Nkomo, J. C. 2005. Energy and Economic Development: Challenges for South Africa. *Journal of Energy in Southern Africa* 16, 10–20.

Oji, C. K. 2015. Bonds: A Viable Alternative for Financing Africa's Development. *SAIIA Occasional Paper* 211, 1–24.

Organisation for Economic Co-operation and Development (OECD). 2015. *Mapping Channels to Mobilise Institutional Investment in Sustainable Energy.* Green Finance and Investment, OECD Publishing. Paris, France.

Reuters. 2015. *World Bank's IFC issues first green bond in South Africa.* Engineering News [Internet]; [updated 2015 November 27]. At: http://www.engineeringnews.co.za/article/world-banks-ifc-issues-first-green-bond-in-south-africa-2015-11-25.

SA Building Review. 2015. *City of Johannesburg — Green Bond* [Internet]; [updated 2015 August 21]. At: http://www.sabuildingreview.co.za/view online/articles/168/advertorials/city-of-johannesburg-green-bond.

Smallridge, D., B. Buchner, C. Trabacchi, M. Netto, J. J. Gomes Lorenzo and L. Serra. 2012. The Role of National Development Banks in Intermediating

International Climate Finance to Scale up Private Sector Investments. *Inter-American Development Bank*, 113.

South African Cities Network (SACN). 2008. *Securing Municipal Finance in Southern Africa.* SACN, 60.

UJUH. 2012. *IDC and PIC Team up in R5 Bn Green Bond* [Internet]; [updated 2015 December 1]. At: http://www.ujuh.co.za/idc-and-pic-team-up-in-r5bn-green-bond/.

United Nations Environment Programme (UNEP). 2015. *Costs of Climate Change Adaptation Expected to Rise Far Beyond Africa's Coping Capacity.* UNEP News Centre [Internet]; [updated 2015 November 28]. At: http://www.unep.org/newscentre/Default.aspx?DocumentID=26788&ArticleID=34788.

Wee, D. 2013. *Going Green? Buy a Kexim bond.* Finance Asia [Internet]; [updated 2013 February 22]. At: http://www.financeasia.com/News/334012,going-green-buy-a-kexim-bond.aspx.

Williams, V. 2014. *Johannesburg Stock Exchange* [Internet]; [updated 2015 August 21]. At: https://www.jse.co.za/news/the-johannesburg-stock-exchange-lists-its-first-green-bond.

World Bank. 2014. *Green Bond Sixth Annual Investor Update 2014.* World Bank Treasury. http://treasury.worldbank.org/cmd/pdf/WorldBankGreenBond Newsletter.pdf.

World Bank. 2015a. *Green Bonds Changing Investor Expectations.* World Bank Feature News Story [Internet]; [updated 2015 January 22]. At: http://www.worldbank. org/en/news/feature/2015/01/22/green-bonds-changing-inves tor-expectations-three-trends.

World Bank. 2015b. *Fact Sheet: Infrastructure in Sub-Saharan Africa.* At: http://web.worldbank.org/WBSITE/EXTERNAL/COUNTRIES/AFRICA EXT/0,,contentMDK:21951811~pagePK:146736~piPK:146830~theSitePK:258644,00.html.

Chapter 6

Critical Natural Resources and Climate Change Mitigation

Anja Brumme

Technische Universität Bergakademie Freiberg,
Faculty of Economics,
especially International Resource Management,
Schloßplatz 1, D-09596 Freiberg, Germany
anja.brumme@vwl.tu-freiberg.de

Abstract

The availability of clean and highly efficient electricity generation technologies is crucial for climate change mitigation. However, in the case of wind power plants, the turbines with the highest conversion efficiency rely on rare earths, i.e., critical natural resources. Failure of the related upstream market might have the potential to impede wind power deployment and, consequently, raise the costs of climate change mitigation.

1. Introduction

Climate change mitigation crucially depends on the abatement of greenhouse gas emissions. Usually, the main focus is on the energy sector where the largest potential for CO_2 emission reductions is perceived [Barrett, 2009]. The availability and cost of carbon-free energy sources is therefore a major determinant for the costs arising from climate change mitigation efforts [Aldy *et al.*, 2010; Weyant, 1993]. However, carbon-free power generation can be associated with additional costs that have, to the author's knowledge, not been considered explicitly in the energy systems modeling literature yet. Namely, renewable energy sources rely on critical natural resources that are traded on highly volatile markets. Distortions on those markets might have the potential to impede the deployment of certain renewable power sources.

In the case of wind power, the most efficient turbines in terms of conversion efficiency require permanent magnets (PM) with rare earth content. However, rare earths have been assessed to be highly critical by agencies all over the world. For example, they appear on the list of critical materials in the United States (US), China and Australia [USDE, 2012] as well as in the European Union (EU) [EC, 2014]. Current scenarios for the development of the energy sector, as adopted in energy system models, usually do not differentiate among subordinate technologies, but regard wind power as a whole. The possibility that natural resources required for the construction of renewable power stations might be unavailable or only available at much higher cost than nowadays is not accounted for. Hence, the estimates of climate change mitigation costs provided by those models might be understated.

A comprehensive overview of critical materials in "green" technologies is given in USDE [2012]. As a representative example, the role of rare earth elements (REE) in wind turbines will be reviewed in more detail in this chapter.[1] The analysis is structured as follows. Section 2 provides a short summary of the technological background of wind power plants. A definition of criticality and overview of material requirements for wind turbines follow in Section 3. Subsequently, a closer look at the market for rare earths is taken in Section 4 since they are identified as the most critical natural resources in wind turbines. Prices on the rare earth market are the subject of Section 5. Policy implications are discussed in Section 6. Section 7 concludes and presents scope for further research.

[1] Although this chapter focuses on wind power, similar problems arise with solar power where Indium, Gallium and Tellurium are required in some applications [USDE, 2012].

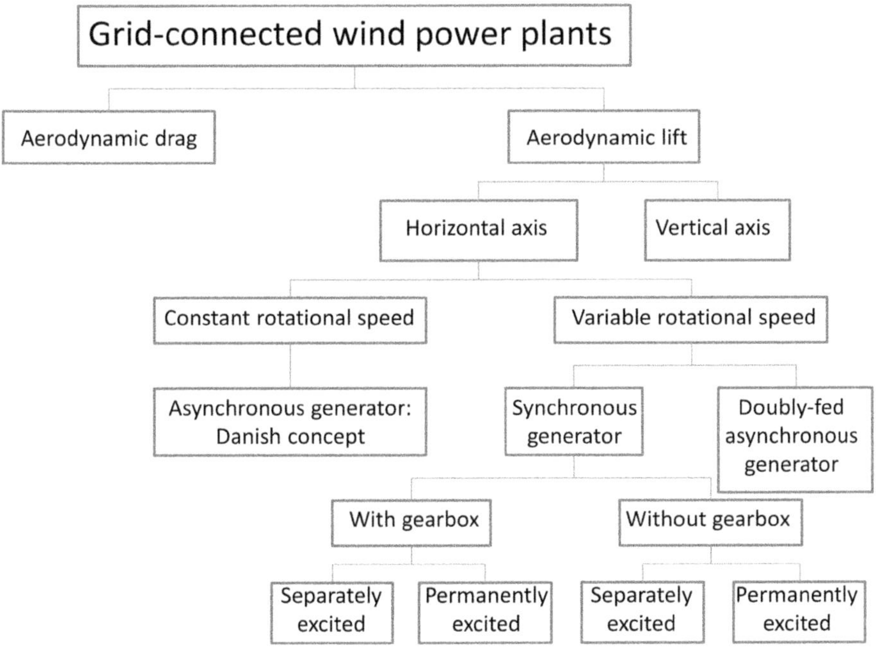

Figure 1. Typification of Wind Power Plants.
Source: Brumme [2014].

2. Wind Power Technologies

Several types of wind turbines are currently in use. Figure 1 provides an overview of the different technical specifications dominating the landscape of grid-connected wind power plants.

The asynchronous generator concept has been prevalent in the 1990s [WWEA, 2006], but has mostly been replaced by now due to the prescriptions of European grid standards [BWE, 2011] and the tendency to build larger-scale wind turbines in the multi-megawatt classes, to which the "Danish concept" is not applicable [Ackermann and Söder, 2002].

Already in 1993, the synchronous generator emerged as an alternative. Its major advantage is the option to omit the gearbox [Heier, 2009], thus enabling a direct-drive concept which has recently gained increased attention, especially for offshore purposes.[2] Direct-drive turbines require low maintenance and are therefore suitable for sites with limited accessibility [Herbert *et al.*, 2007]. Moreover, they exhibit the largest conversion efficiency if operated with a permanent magnet generator. Instead of being permanently excited, they may also be separately excited, i.e., utilize an electromagnet, yielding a somewhat smaller efficiency degree in the range of that of doubly-fed asynchronous generators, the third main type [Hau, 2008].

Doubly-fed asynchronous generators nowadays still dominate the scene, but they are progressively replaced with gearless designs: the latter comprised a market share of 28% of total wind power capacity in 2013, rising up from 12% in 2008 [REN21, 2014]. However, the doubly-fed asynchronous generator is a simple, low-cost variant applicable to high power wind turbines [BWE, 2011; WWEA, 2006].

Based on this background information, the analysis now proceeds to the issue of critical material requirements in wind power plants.

3. Critical Materials in Wind Turbines

Referring to the definition of the EC [2014], which is also a much-cited consensus in the literature, two aspects have to be considered for the evaluation of criticality: economic importance, i.e., demand-side subjects, and supply risk, with a special focus on poor governance. The materials integrated in the EC's analysis and their assessment with respect to the two dimensions of criticality are shown in Figure 2. A natural resource is classified as critical if the criticality threshold is exceeded for both dimensions, indicated by the shaded area in the top right corner of the graph.

In the next step, a review of material intensities required for the construction of wind turbines is provided for comparison with the EC's list of critical materials. Data from the life cycle analysis literature, however, are quite scarce and differ considerably, as

[2]Onshore potentials for wind power have largely been exploited already, so further wind power deployment will mainly concentrate on offshore locations [Bilgili *et al.*, 2011].

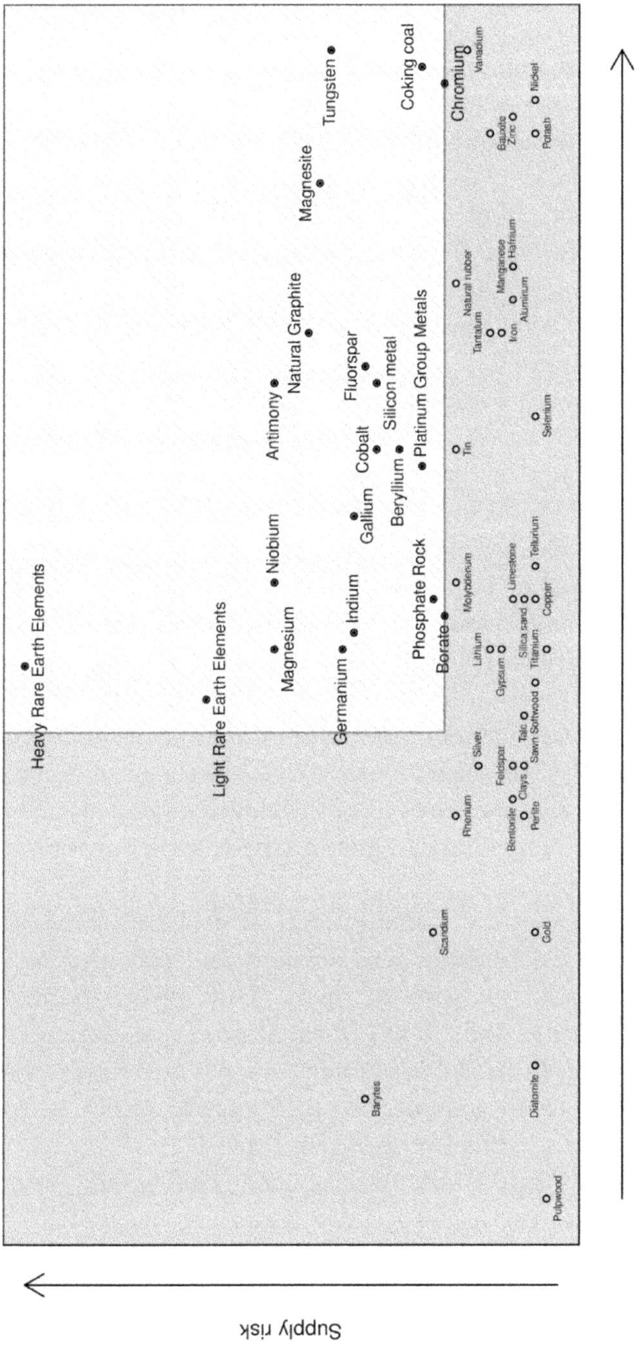

Figure 2. Criticality Assessments for the EU for 2013.

Source: EC [2014].

Table 1 shows. This is partly due to uncertainty associated with the non-disclosure of manufacturer data, but partly also originates from the different scope of the studies. Jacobson and Delucchi [2011] incline to generalization, but only list the materials without giving quantitative information, except for magnetic materials that they place in the range of about 0.2 kg/kW of rated capacity. Ardente *et al.* [2008] indicate data for an Italian wind farm with 660 kW turbines. Martínez *et al.* [2009] refer to a 2 MW power plant located in Spain. Schleisner [2000] investigates Danish turbines and differentiates among an onshore and an offshore wind farm. Kleijn and Voet [2010] also make that distinction and focus on an 800 kW onshore and a 2 MW offshore power plant. Weinzettel *et al.* [2009] study a 5 MW offshore floating turbine produced in Norway.

As Jacobson and Delucchi [2011] point out, magnetic materials are identified as critical because of their rare earth content. When comparing the materials listed in Table 1 with the critical resources identified in Figure 2, it becomes obvious that rare earths are the only materials that appear to be critical, which justifies a closer look at them.

As already mentioned in Section 2, magnetic materials are exclusively needed for wind turbines operated with a synchronous generator, which explains their selective occurrence in the studies cited in Table 1. Moreover, manufacturers have the choice among electromagnets and permanent magnets, whereby rare earths are only employed in the latter.

A more recent study by Zimmermann *et al.* [2013] takes those different types of wind turbines into account and provides detailed material intensity data for each of them. They give a permanent magnet amount of about 0.6 kg/kW of rated power for both onshore and offshore applications using permanent magnet generators. 30% of that amount is rare earth content, i.e., 0.18 kg/kW, which is closely in line with previous studies like those in Table 1.

Since supply risk appears to be the most challenging issue for the provision of rare earths, especially regarding heavy REE (see Figure 2), supply-side market failure is discussed in detail in the following section.

Table 1. Material Intensities for Wind Turbines.

Material	Jacobson and Delucchi [2011]	Ardente *et al.* [2008]	Martínez *et al.* [2009]	Schleisner [2000]		Kleijn and Voet [2010]		Weinzettel *et al.* [2009]
				onshore	offshore	onshore	offshore	offshore
Concrete	x		350.00	565.00	1,130.00	306.00	1,050.00	
Steel	x	100.60	89.90	105.40	113.20	135.00	148.00	311.00
Sand, gravel			0.20	4.20	4.20	0.00	150.00	646.00
Iron		9.10	28.80	24.00	48.00			13.80
Glass fiber	x	7.50	4.40					10.40
Oil products		0.20		0.20	0.20			15.00
Copper	x	1.40	1.80	0.70	5.90	1.83	2.75	11.70
Wood epoxy, resin	x		6.60					0.30
Lead					6.70	0.00	3.79	2.60
Plastic	x	0.10		4.00	4.00	1.92	2.09	1.80
Polyethylene					1.10			2.90
Glass, ceramics				2.20	2.20			0.01
Aluminium	x	0.10		2.80	2.80	0.26	0.42	0.50
Rare earths	0.20						0.15	
Logs (wood)								0.07
Others		0.60		1.40	1.40			0.05

Note: Numbers have been normalized to kg per kW of rated capacity.
Source: Adapted from Brumme [2014].

4. Supply-Side Market Failure

The materials under consideration in this chapter are shortly presented in the subsequent box to ensure traceability of the argument for the non-expert reader.

In the following, three kinds of distortions on the REE market that may preclude the establishment of an efficient price mechanism will be explained in more detail: market power, which is closely related to governance issues; negative externalities; and joint production. As mentioned before, the focus will be on the supply side.

4.1. *Market power and governance issues*

Along the REE supply chain, concentration is perceptible to a different extent and on different levels. The stages of mining, processing, fabrication of components and fabrication of end-use technologies (with a special focus on PM in wind turbines) are to be considered here.

4.1.1. *Mining*

Concerning the extraction process, Figure 3 gives an impressive idea of the current situation on country level with China supplying almost the whole amount of REE on the world market. This outstanding position has been attained in the 1980s, when China offered REE for a lower price than its competitors, whereby companies from other countries were successively crowded out [Haxel *et al.*, 2002]. Once China had established as a market leader, export restrictions and tariffs have resulted in artificial shortages on the world market and severe price volatility [USDE, 2012].

Box 1: Rare Earth Elements

Rare Earth Elements (REE) are a group of chemical elements with similar properties. They comprise the 15 lanthanides (atomic numbers 57 through 71), scandium (21) and yttrium (39). A subdivision is made into light REE, i.e., the elements with atomic

Box 1: (*Continued*)

numbers 57 to 63, and heavy REE, ranging from atomic numbers 64 to 71, plus yttrium. Geologically they occur as compounds, whereby the concentration of light REE is usually higher than that of heavy REE [Walters and Lusty, 2011].

The geological concentration of REE in total is moderate with a Herfindahl-Hirschman Index value of 0.23 [Angerer *et al.*, 2009]. Although only about 40% of reserves are located in China, more than 90% of REE currently mined originate from there [Binnemans *et al.*, 2013; Pothen, 2014]. China has been able to attain such a dominating position due to extraordinarily low production costs, made possible by low energy and labor costs and a lack of environmental standards [Gavin, 2013].

REE are mostly applied in high-technology industries. Besides permanent magnets, they are also an important component of phosphors in liquid crystal displays (LCD) and energy-saving light bulbs, catalysts in chemical processing, and batteries [Brumme, 2014].

For permanent magnets (PM) employed in wind power plants, the following five elements are required [Goonan, 2011]:

Element	Atomic number	Consumption in PM (share of total demand)
Praseodymium	59	70%
Neodymium	60	76%
Gadolinium	64	69%
Terbium	65	11%
Dysprosium	66	100%

As the data indicate, the PM industry consumes a significant share of most relevant REE materials. Therefore, any market distortions with respect to REE provision, either in terms of extraordinary price volatility or even in terms of physical shortages, can be expected to exert a non-negligible influence on the wind power industry and, consequently, on the cost of climate change mitigation.

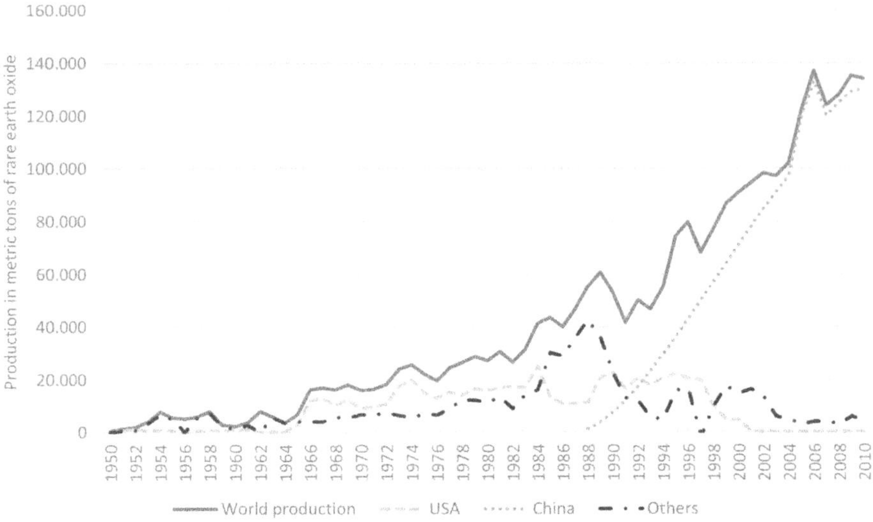

Figure 3. REE Mine Production since 1950 by Country.
Sources: Brumme [2014], Kelly and Matos [2014], and Wang *et al.* [2015].

However, REE production at other sites could not be started or re-started immediately since mining requires substantial monetary investment, planning and legal permission processes and possibly the set-up of an adequate infrastructure [EC, 2008]. Thus, Chinese predominance could be retained until now.

That fact raises the question of governance issues which has been indicated in the previous section. According to the Worldwide Governance Indicator provided by the World Bank, China reveals a negative governance score in all categories,[3] resulting in the country being ranked 36th percentile if all categories are weighted equally [The World Bank Group, 2014]. It can be concluded that it might not be a very reliable supplier, which proved to be true: for some elements, such as Neodymium, Terbium, Dysprosium and Europium, demand

[3]The categories are Voice and Accountability, Political Stability and Absence of Violence, Government Effectiveness, Regulatory Quality, Rule of Law, and Control of Corruption. Values are indicated on a scale from −2.5 to +2.5.

is estimated to have exceeded supply already, due to the creation of artificial shortages [Cordier, 2011].

4.1.2. *Processing and fabrication*

On the following stage, refining and processing of rare earth metals and oxides is also politically encouraged to be carried out in China [Liedtke and Elsner, 2009]. Semi-fabrication of REE applications mainly takes place in China as well. In contrast to mineral extraction where foreign investment is confined, the settlement of foreign companies in China is requested in subsequent steps along the supply chain. End-use fabrication is more widely distributed internationally [Goonan, 2011]. Still, in the case of Neodymium-Iron-Boron permanent magnets, manufacturing is restricted to 10 companies in China, Japan and Germany due to intellectual property rights [USDE, 2012]. Wind turbine production is, finally, more widespread. About 20% of wind power plants were manufactured in China in 2013, a market share comparable to that of Germany [REN21, 2014].

4.2. *Negative externalities*

Negative external effects mainly arise at the REE extraction and refining stages and are particularly related to environmental damage and the respective consequences for humans. For instance, toxic chemicals are utilized during the whole mining, separation and refinement procedure. Since proper waste disposal is often neglected, especially at illegally operated mines in China, toxic sludge finally enters the ground water which leads to pollution, contamination of agricultural land, and health impairment of the local population. Moreover, the high energy intensity of REE production contributes to air pollution and global warming as electricity in China is still generated in coal-fired power plants to a large extent [Lohse, 2011; Walters and Lusty, 2011].

In the past, radioactivity was another important issue connected with REE mining. When REE were extracted from monazite minerals, nuclear radiation was released. As a consequence, monazite is rarely exploited any more [Haxel *et al.*, 2002].

With respect to environmental issues, there is another indicator for international ranking of countries, similar to the Worldwide Governance Indicator. The Environmental Performance Index, published by the Yale Center for Environmental Law and Policy at Yale University, evaluates 180 countries with respect to nine categories.[4] In 2016, China is ranked 109th, with a score of 65.1 out of 100 points. Compared to the previous evaluation in 2014 when China gained only 43 points and was ranked 118th out of 178, a considerable improvement has been made. This is also reflected in the positive intertemporal trend: within the last 10 years China's score increased by almost 13% [Yale University, 2014; Yale University, 2016].

4.3. *Joint production*

REE exhibit joint production properties in two ways: first, as specialties, they are usually only by-products of commodities that are rather mined and used as bulk materials. Second, they occur as compounds, which means that they have to be extracted as a bundle, irrespective of actual demand patterns. More precisely, Neodymium is often the main target of REE mining, but other REE can only be separated from it at a later stage, so they are extracted as well, which has resulted in excess supply for some elements [Angerer *et al.*, 2009].

The following section sheds some light on the implications of those special REE market properties for pricing.

5. Consequences for REE Price Formation

The actual REE price formation process is rather intransparent since a centralized platform for spot trading has only been established in 2013 [Shen, 2013]. Before, bilateral contracts with fixed prices for a certain period of time were prevalent [Behrendt *et al.*, 2007; USDE,

[4]These categories are Health Impacts, Air Quality, Water and Sanitation, Water Resources, Agriculture, Forests, Fisheries, Biodiversity and Habitat, and Climate and Energy.

Table 2. Rare Earth Oxide Prices in USD/kg for Selected Elements from 2004 to 2010 (Referring to a Minimum Purity of 99%).

Rare earth oxide	2004	2007	2008	2009	2010
Praseodymium	7.44	28.00	27.00	14.00	72.00
Neodymium	5.65	29.00	27.00	14.00	77.00
Terbium	341.00	555.00	560.00	350.00	605.00
Dysprosium	31.00	85.00	110.00	100.00	295.00

Source: ISE [2014].

2012], specialized REE trading companies used to act as intermediaries between raw materials suppliers and processing firms [Walters and Lusty, 2011].

In general, price formation of REE underlies many influences and distortions, as described in the previous section, precluding the achievement of a price mechanism that would yield an efficient market outcome. The standard theoretical concept for the optimal intertemporal depletion path of a non-renewable natural resource, as prescribed by Hotelling's rule, requires the existence of a perfectly competitive market [Hotelling, 1931]. However, Chinese market power resulted in quantitative restrictions as well as price volatility that could be observed in the past as shown by Table 2 for the materials occurring in permanent magnets.[5]

Basic economic theory predicts that the monopolist (or, in this case, a quasi-monopolist) will offer a smaller quantity for a higher price than under the circumstances of perfect competition. Yet, the existence of negative externalities as a second aspect of market failure interferes with market power in such a way that, a priori, it cannot be determined if actual supply will be higher or lower than the social optimum because the internalization of those externalities requires a quantity reduction and price increase. In a recent study by Pothen [2014], welfare effects are calculated when comparing actually observed REE prices to the outcome of a market model under perfect competition. Still, his analysis is restricted to the aspect of market

[5] Data for Gadolinium were not available.

power as he explicitly excludes environmental effects. An estimate of the overall effect which includes both market power and externalities has, to the author's knowledge, not been provided yet.

With respect to price formation, joint production also plays a crucial role. It yields a low price elasticity of supply for single REE in the short run, i.e., a relatively steep supply curve. Combined with a low price elasticity of demand, resulting from limited substitution possibilities in the production process of final products, the main effect of exogenous shocks arises as a price adjustment rather than a quantitative reaction, which explains price volatility.

For end-use applications such as wind turbines, REE price volatility might be of minor concern since they are used in relatively small quantities (cp. Table 1). However, physical shortages have already been observed for materials in permanent magnets, namely Neodymium, Terbium and Dysprosium [Cordier, 2011]. Thus, wind energy deployment might be affected such that the most efficient turbines that are, moreover, preferred for offshore sites, are not available to a sufficient amount. Hence, they have to be substituted by alternative electricity generation technologies, probably other wind turbine concepts that are less efficient. If they are replaced by nonrenewable sources, the compliance with greenhouse gas emission targets is impeded. In both cases, climate change mitigation costs can be expected to rise.

6. Policy Implications

REE export restrictions have already been imposed by China since 1999, but did not become binding before 2010. Their persistent maintenance resulted in a formal complaint at the World Trade Organization (WTO). The EU, Japan and the US argued that China violated both general WTO directives and its particular accession commitments [Gavin, 2013]. As mentioned previously, the occurrence of market power and negative external effects operate in opposite directions. Therefore, China justified tariffs and export quotas on rare earths as protective measures for the environment and contribution to resource conservation. The fact that restrictions only applied to exports, not production volumes, so that the domestic industry benefited from a competitive advantage in raw materials provision, sheds some doubt on this argument. Finally, the WTO obliged China to withdraw its export restrictions and in December 2014 China announced to abolish the trade barriers [Paul, 2015].

The environmental problems connected to REE extraction have been recognized and tackled by the Chinese government. In 2011, new environmental standards and specific emission limits for 15 pollutants were passed [Lohse, 2011]. Furthermore, several illegally operated mines have been closed [Gavin, 2013]. However, the environmental and social conditions in and around Chinese mines are still reported to be inadequate[6] [Kaiman, 2014]. Although it is always difficult to negotiate and enforce uniform international standards with respect to both human rights and environmental protection [Barrett, 2013], efforts have been made recently to develop corporate social responsibility indicators for the mining sector [e.g., Azapagic, 2004] and an increasing number of mining companies monitor their performance in accordance with those or similar means [Jenkins and Yakovleva, 2006].

In order to ensure security of supply for import-dependent countries, different measures have been taken by their governments. They include the creation of national stockpiles, continuous monitoring of developments on raw materials markets, or bilateral trade agreements. Moreover, funding for domestic exploration and mining activities and for research in the fields of critical materials recycling, substitution and material efficiency is common [USDE, 2012].

The development of mining sites outside China was pursued intensely especially after a sharp increase of rare earth prices in 2011. Since then, however, prices have fallen significantly [USGS, 2015], forcing some companies like Molycorp, operator of the Mountain Pass Mine in the US, to refrain from production [Molycorp, 2015].

---

[6]Although very few studies refer to social conditions in the mining sector, the protection of human rights is still one of the most important global public goods and should not be omitted in the discussion [Barrett, 2013].

7. Summary and Outlook

It can be expected that permanently excited direct-drive wind turbines will gain importance in the future due to the increased development of offshore sites. Already in 2010, the majority of the 10 global market-leading wind turbine producers employed permanent magnet generators or was at least investigating the technology [Kurronen *et al.*, 2010]. Therefore, failure of the upstream market for critical natural resources has the potential to preclude a first-best solution for highly efficient electricity provision under greenhouse gas emission constraints, thereby increasing climate change mitigation costs. The case of rare earth elements in wind power plants has received little attention so far, but entails three main problems: Chinese market power, negative externalities, and joint production. As a result, high price volatility and physical shortages arose on the REE market and are expected to remain in the short to medium term until new mines outside China take and consolidate their position in the market or recycling and substitution options become marketable.

Scope for further research in economic terms is mainly seen in three fields. First, an estimate of overall welfare effects of the REE mining process, taking into account both market power and externality aspects, is still lacking. Such an analysis could even be extended to include the complete life cycle of a wind turbine, explicitly weighing the impacts of natural resource extraction against its contribution to clean electricity provision. Second, a tangible numerical estimate for the additional costs of climate change mitigation could be obtained by integrating a natural resource constraint in renewable energy production functions within an energy-economy-climate model. Third, incentive-compatible mechanisms for the implementation of environmental and social standards in the mining industry should be investigated in more detail.

References

Ackermann, T. and L. Söder. 2002. An Overview of Wind Energy-Status 2002. *Renewable and Sustainable Energy Reviews* 6 (1–2), 67–128.

Aldy, J. E., A. J. Krupnick, R. G. Newell, I. W. H. Parry and W. A. Pizer. 2010. Designing Climate Mitigation Policy. *Journal of Economic Literature* 48 (4), 903–934.

Angerer, G., F. Marscheider-Weidemann, A. Lüllmann, L. Erdmann, M. Scharp, V. Handke and M. Marwede. 2009. *Schlussbericht Rohstoffe für Zukunftstechnologien. Einfluss des Branchenspezifischen Rohstoffbedarfs in Rohstoffintensive Zukunftstechnologien auf die Zukünftige Rohstoffnachfrage.* 2nd edn. Fraunhofer IRB Verlag. Stuttgart, Germany.

Ardente, F., M. Beccali, M. Cellura and V. L. Brano. 2008. Energy Performances and Life Cycle Assessment of an Italian Wind Farm. *Renewable and Sustainable Energy Reviews* 12 (1), 200–217.

Azapagic, A. 2004. Developing a Framework for Sustainable Development Indicators for the Mining and Minerals Industry. *Journal of Cleaner Production* 12 (6), 639–662.

Barrett, S. 2009. The Coming Global Climate-Technology Revolution, *Journal of Economic Perspectives* 23 (2), 53–75.

Barrett, S. 2013. *Why Cooperate? The Incentive to Supply Global Public Goods.* Oxford, UK: Oxford University Press.

Behrendt, S., M. Scharp, W. Kahlenborn, M. Feil, C. Dereje, R. Bleischwitz and R. Delzeit. 2007. *Seltene Metalle. Maßnahmen und Konzepte zur Lösung des Problems konfliktverschärfender Rohstoffausbeutung am Beispiel Coltan.* Texte 08-07. Umweltbundesamt (ed.), Dessau, Germany.

Bilgili, M., A. Yasar and E. Simsek. 2011. Offshore Wind Power Deployment in Europe and Its Comparison with Onshore Counterpart, *Renewable and Sustainable Energy Reviews* 15 (2), 905–915.

Binnemans, K., P. T. Jones, B. Blanpain, T. V. Gerven, Y. Yang, A. Walton and M. Buchert. 2013. Recycling of Rare Earths: A Critical Review. *Journal of Cleaner Production* 51, 1–22.

Brumme, A. 2014. *Wind Energy Deployment and the Relevance of Rare Earths: An Economic Analysis.* Springer Gabler (BestMasters). Wiesbaden, Germany.

BWE. 2011. *Technik. Kapitel 5: Konstruktiver Aufbau einer Windkraftanlage.* German Wind Energy Association. At: http://www.wind-energie.de/infocenter/technik.

Cordier, D. J. 2011. *2009 Minerals Yearbook, Volume I: Metals and Minerals.* U.S. Geological Survey (ed.), Reston, Virginia.

EC. 2008. *The Raw Materials Initiative: Meeting Our Critical Needs for Growth and Jobs in Europe.* Communication from the Commission to the European Parliament and the Council. Brussels, Belgium.

EC. 2014. *Report on Critical Raw Materials for the EU: Report of the Ad Hoc Working Group on Defining Critical Raw Materials.* European Commission Enterprise and Industry Directorate General. Brussels, Belgium.

Gavin, B. 2013. China's Growing Conflict with the WTO: The Case of Export Restrictions on Rare Earth Resources. *Intereconomics* 48 (4), 254–261.

Goonan, T. G. 2011. *Rare Earth Elements: End Use and Recyclability.* U.S. Department of the Interior and U.S. Geological Survey (eds.), Reston, Virginia.

Hau, E. 2008. *Windkraftanlagen. Grundlagen, Technik, Einsatz, Wirtschaftlichkeit*, 4th edn. Springer. Berlin and Heidelberg, Germany.

Haxel, G. B., J. B. Hedrick and G. J. Orris. 2002. *Rare Earth Elements — Critical Resources for High Technology.* USGS Fact Sheet 087-02. U.S. Geological Survey (ed.), Reston, Virginia.

Heier, S. 2009. *Windkraftanlagen: Systemauslegung, Netzintegration und Regelung*, 5th edn. Vieweg + Teubner, Wiesbaden, Germany.

Herbert, G. M. J., S. Iniyan, E. Sreevalsan and S. Rajapandian. 2007. A Review of Wind Energy Technologies. *Renewable and Sustainable Energy Reviews* 11 (6), 1117–1145.

Hotelling, H. 1931. The Economics of Exhaustible Resources. *Journal of Political Economy* 39 (2), 137–175.

ISE. 2014. *Aktuelle und historische Marktpreise der gängigsten Seltenen Erden* [Internet]; [updated 2016 May 13]. At: http://institut-seltene-erden.org/aktuelle-und-historische-marktpreise-der-gangigsten-seltenen-erden/.

Jacobson, M. Z. and M. A. Delucchi. 2011. Providing All Global Energy with Wind, Water, and Solar Power, Part I. Technologies, Energy Resources, Quantities and Areas of Infrastructure, and Materials, *Energy Policy* 39 (3), 1154–1169.

Jenkins, H. and N. Yakovleva. 2006. Corporate Social Responsibility in the Mining Industry: Exploring Trends in Social and Environmental Disclosure. *Journal of Cleaner Production* 14 (3–4), 271–284.

Kaiman, J. 2014. *Rare Earth Mining in China: The Bleak Social and Environmental Costs* [Internet]; [updated 2014 March 20]. At: http://www.theguardian.com/sustainable-business/rare-earth-mining-china-social-environmental-costs.

Kelly, T. D. and G. R. Matos. 2014. *Historical Statistics for Mineral and Material Commodities in the United States* [Internet]; [updated 2016 March 30]. At: http://minerals.usgs.gov/minerals/pubs/historical-statistics/.

Kleijn, R. and E. van der Voet. 2010. Resource Constraints in a Hydrogen Economy Based on Renewable Energy Sources: An Exploration. *Renewable and Sustainable Energy Reviews*, 14 (9), 2784–2795.

Kurronen, P., M. Haavisto and J. Pyrhönen. 2010. *Challenges in Applying Permanent Magnet (PM) Technology to Wind Power Generators* [Internet]; [updated 2013]. At: http://www.theswitch.com/wp/wp-content/uploads/2013/10/EWEC-2010_Paper_Challenges-in-applying-PM-technology_Panu-Kurronen-The-Switch_final-2.0_100417.pdf.

Liedtke, M. and H. Elsner. 2009. *Seltene Erden.* Commodity Top News 31. BGR (ed.), Hannover, Germany.

Lohse, J. 2011. *Umweltschäden durch Neodym in der Windkraft* [Internet]; [updated 2011 May 27]. At: http://www.cleanenergy-project.de/erneuerbare-energien/4301-seltene-erden-in-der-windkraft?showall=&Itemid=164.

Martínez, E., F. Sanz, S. Pellegrini, E. Jiménez and J. Blanco. 2009. Life Cycle Assessment of a Multi-Megawatt Wind Turbine. *Renewable Energy* 34 (3), 667–673.

Molycorp. 2015. *Molycorp to Move Its Mountain Pass Rare Earth Facility to "Care And Maintenance" Mode* [Internet]; [updated 2015 August 26]. At: http://www.molycorp.com/molycorp-to-move-its-mountain-pass-rare-earth-facility-to-care-and-maintenance-mode/.

Paul, S. 2015. *China's Rare Earths Quotas Go, Possible New Moves Stoke Supply Doubts* [Internet]; [updated 2015 January 7]. At: http://www.reuters.com/article/2015/01/07/china-rareearths-producers-idUSL3N0UL65220150107.

Pothen, F. 2014. Dynamic Market Power in an Exhaustible Resource Industry: The Case of Rare Earth Elements. *ZEW Discussion Paper 14-005.*

Renewable Energy Policy Network for the 21st Century (REN21). 2014. *Renewables 2014 Global Status Report.* REN21. Paris, France.

Schleisner, L. 2000. Life Cycle Assessment of a Wind Farm and Related Externalities. *Renewable Energy*, 20 (3), 279–288.

Shen, H. 2013. China's First Rare Earth Spot Trading Platform to Stabilize Rare Earth Prices. Toronto, Canada. At: http://investorintel.com/rare-earth-intel/chinas-first-rare-earth-spot-trading-platform-put-test/.

The World Bank Group. 2014. *Worldwide Governance Indicators.* Country Data Reports. Washington, D.C. At: http://info.worldbank.org/governance/wgi/index.aspx#reports.

U.S. Department of Energy (USDE). 2012. *Critical Materials Strategy.* DOE/PI-0009. USDE (ed.), Washington, D.C.

U.S. Geological Survey (USGS). 2015. *Mineral Commodity Summaries* [Internet]; [updated 2015 January 30]. At: http://minerals.usgs.gov/minerals/pubs/mcs/2015/mcs2015.pdf.

Walters, A. and P. Lusty. 2011. *Mineral Profile: Rare Earth Elements* [Internet]; [updated 2011 November]. At: http://www.bgs.ac.uk/mineralsuk/search/home.html.

Wang, X., Y. Lei, J. Ge and S. Wu. 2015. Production Forecast of China's Rare Earths Based on the Generalized Weng Model and Policy Recommendations. *Resources Policy* 43, 11–18.

Weinzettel, J., M. Reenaas, C. Solli and E. G. Hertwich. 2009. Life Cycle Assessment of a Floating Offshore Wind Turbine. *Renewable Energy* 34 (3), 742–747.

Weyant, J. P. 1993. Costs of Reducing Global Carbon Emissions, *Journal of Economic Perspectives* 7 (4), 27–46.

World Wind Energy Association (WWEA). 2006. *Wind Energy — Technology and Planning* [Internet]; [updated 2006]. At: http://www.wwindea.org/technology/ch01/estructura-en.htm.

Yale University. 2014. *2014 EPI Framework +Indicator Scores* [Internet]; [updated 2014 January]. At: http://epi.yale.edu/downloads.

Yale University. 2016. *Global Metrics for the Environment: 2016 Report* [Internet]; [updated 2016]. At: http://epi.yale.edu/sites/default/files/2016EPI_Full_Report.pdf.

Zimmermann, T., M. Rehberger and S. Gößling-Reisemann. 2013. Material Flows Resulting from Large Scale Deployment of Wind Energy in Germany. *Resources* 2 (3), 303–334.

Part 2

Theoretical Perspective

Chapter 7

Putting Climate Finance into Context: A Global Public Goods Perspective

Inge Kaul

Hertie School of Governance,
Quartier 110, Friedrichstrasse 180, 10117 Berlin, Germany
kaul@hertie-school.org and inge-kaul@t-online.de

Abstract

The chapter explores whether there is added value in examining the financing of climate change mitigation and adaptation through the analytical lens of global public goods (GPGs). The discussion indicates that the answer to this question is "yes". The GPG lens brings into focus the current haphazard nature of climate finance, which often appears to happen without clear criteria for determining when, where and why to use which type of finance. Importantly, efficiency and equity considerations are often confounded to the detriment of both. The analysis suggests that, in order to approach climate finance more systematically and effectively, it would need to be placed into context: into a broader theory of global public economics and finance, of which we have, at present, only some incipient elements. The analysis also indicates, for further research and debate, what could be done to correct this situation.

1. Introduction

In *The Economics of Climate Change*, Stern [2007: 1] refers to atmospheric quality as a global public good (GPG) and to climate change as "the greatest example of market failure we have ever seen". Amid the burgeoning social-science literature on climate finance, however, studies that use the GPG concept as their analytical lens remain rare.

This chapter explores whether considering the GPG properties of climate change provides added value when examining issues of climate finance.[1] Could the GPG lens improve our understanding of climate finance? More importantly, would it allow us to discover new ways to meet the challenge of climate change by fostering global sustainable growth and development, as stipulated in Agenda 2030 and the Paris Agreement?[2]

These questions are explored below in two stages. Section 2 introduces the GPG concept, highlighting some of the ways GPGs differ from national public goods (NPGs), as those could indicate

---

[1]The 2014 *Biennial Assessment and Overview of Climate Finance Flows Report* of the Standing *Committee on Finance* (SCF) of the United Nations Framework Convention on Climate Change (UNFCCC) suggests to define climate finance as "including all financial flows whose expected effect is to reduce net GHG [greenhouse gas] emissions and/or to enhance resilience to the impacts of climate variability and the projected climate change" [UNFCCC/SCF, 2014: 15]. As the SCF notes, climate finance may include private and public, domestic and international flows. Its reports address both, "global total climate finance" and "climate finance from developed to developing *cou*ntries" [UNFCCC/SCF, 2014: 15–16].

[2]Agenda 2030 was adopted in September 2015 by all 193 Member States of the United Nations. It sets forth 17 main goals that the international community decided to pursue over the next 15 years to end extreme poverty, fight inequality and injustice, and protect the planet. For the full text of Agenda 2030 see http://www.un.org/ga/search/view_doc.asp?symbol=A/RES/70/1&Lang=E/. *The Paris Agreement* was adopted in December of the same year. It is the outcome document of the 21st session of the UNFCCC. In its opening paragraphs, the Paris Agreement emphasizes "the intrinsic relationship that climate change actions, responses and impacts have with equitable access to sustainable development and eradication of poverty". (See document UNFCCC/CP/2015/L.9/Rev. 1 of 12/12/2015: 21 available at http://unfccc.int/files/essential_background/convention/application/pdf/english_paris_agreement.pdf/). Clearly, the intrinsic relationship between climate change mitigation and adaption, on the one hand, and poverty eradication, on the other hand, which Agenda 2030 and the Paris Agreement emphasize, has implications for how the challenge of climate change would need to be addressed, as we will discuss later, in Section 3.4.

possible inhibitors and facilitators of climate finance. Against this background, Section 3 presents four issues that come into focus and appear in a new light when examining climate finance through the lens of GPGs: (1) the differences and synergies between climate finance and development finance; (2) the risk of market and state failure in pursuing and financing climate change mitigation; (3) the importance of intra-generational fairness; and (4) the current lack of integrated provision-path analysis and management.

Two main findings emerge from the analysis. First, international public climate finance (IPCF) is an important inhibitor of overall climate finance.[3] Second, many of the factors constraining climate finance are not climate-specific but rather GPG-specific. They affect GPGs across the board and reflect a basic problem of institutional lock-in, whereby national and international governance systems have not yet been adjusted to address GPG-type policy challenges efficiently and effectively.

The final section thus concludes that enhancing climate finance might require a two-pronged approach — developing the incipient effort in this chapter to apply the GPG lens to climate issues while building a general theory of GPG provision and finance, of which only the initial stage is underway.

Therefore, the task of clarifying GPG characteristics is taken up next.

2. Introducing Global Public Goods

Scholars commonly define a GPG simply as a good that differs in geographic and/or temporal reach from an NPG, the type of public good (PG) on which conventional public economic theory focuses. Such a definition misses two important points, however. First, as Desai [2003] argues, we cannot take the generic PG definition of public goods as settled; second, the wider reach of GPGs' benefits and costs is but one of several dimensions in which NPGs and GPGs differ.

[3] The term "international public climate finance" refers to public revenue channeled to bilateral or multilateral climate-related programs and projects.

Moreover, as the discussion below makes clear, the other neglected features appear to be precisely those that might explain why many GPGs, including climate change, are allowed to linger for so long in well-documented states of worsening underprovision, even though possible courses of corrective action are known but adopted only hesitantly.

It is therefore useful to, first, discuss the generic PG concept and, second, identify the key distinguishing features of GPGs.[4]

2.1. *The generic definition of public goods*

According to standard economic theory, PGs are marked by non-rivalry and/or non-excludability in consumption. When possessing both these properties, they are said to be "pure public in consumption"; when exhibiting only one such characteristic, they are described as "impure public in consumption". Goods that are fully or partially rival and/or excludable are referred to as "pure private" or "impure private" goods.

Besides agreeing on these basic elements of the definition of PG, scholars also agree that, generally, a good's publicness or privateness in consumption is not an innate property but a social construct — the result of a socio-cultural or political choice. However, they differ about how impure public a good can be and still be considered public and which of the two properties (non-rivalry or non-excludability) matters most in classifying a good as public. Another open question is whether to attach a value connotation to PGs. While some scholars, including textbook authors, still refer to PGs as "beneficial" or "to be enjoyed by all", others take a more empirical, value-neutral position, recognizing that, as people's living conditions tend to vary widely, their preferences for certain PGs are also likely to vary.

The definition of generic PG used in this chapter is the value-neutral two-tier definition presented in Box 1. This definition distinguishes between a good's potential for being public in consumption

[4]For a more detailed discussion on these different types of PG, see, among others, Kaul *et al.* [2003] and Kaul *et al.* [2016].

(i.e., available for all or for anyone to be affected by it) and its actual publicness (i.e., being de facto non-excludable or non-exclusive).

Box 1: Defining Public Goods

Public goods are goods whose properties are non-rival in consumption, non-excludable, or both. If a good has both properties, it is said to be "pure public"; if it has just one of these characteristics, it is referred to as "impure public".

Thus, public goods are defined as opposites of private goods, which, according to standard economic theory, are rival and excludable.

However, being public or private is, in many cases, not an innate property of the good but a social construct, reflecting either a deliberate social choice or inaction due to a lack of scientific knowledge or information about the good.
Accordingly, it seems useful to distinguish between goods that have the potential to be public in consumption and those that are actually in the public domain, as does the following two-tier definition:

Definition 1: *Goods have a special potential for being public if they are non-excludable and/or non-rival in consumption.*

Definition 2: *Goods are de facto public if they are non-exclusive and available for all.*

The advantage of this definition is that it encourages analysts to ask whether and according to which criteria it is socially desirable for non-rival goods not to be fully or, at least, partially non-exclusive, for fully or partially rival goods to be non-excludable, and for potentially excludable goods to be left or placed in the public domain.

Source: This box draws on Kaul and Mendoza [2003].

Conventionally, PGs have been considered mainly from their consumption side. Due to their publicness in consumption, they have been seen as one of the conditions that cause market failure and may thus justify state intervention in the economy.[5] However, this conclusion rests on the assumption that individual actors are selfish and rational and would therefore free-ride in the presence of a public good (i.e., let someone else provide the good and enjoy it for free). As a result, PGs became an integral component of the theory of the state and the economics of the public sector. For a long time, only limited attention was paid to their production.[6] The state was expected to take care of that and did, in fact, play a relatively significant role in PG production from the 1950s to the 1970s, when the foundations of conventional public economics and, as one of its core elements, those of conventional PG theory were laid.[7]

Advancing privatization and economic liberalization has turned PGs into multi-actors, sometimes even voluntarily or privately provided products. It is now quite common for scholars to differentiate among PGs according to their consumption properties and their "social aggregation technology", depending on whether their production is a summation, weakest-link, or best-shot process.[8] Increasingly, scholars not only consider the overall supply technology of a good but

[5]The conditions that make markets fail usually include (besides public conditions) imperfect competition, externalities, incomplete markets, imperfect information, and macroeconomic disturbances [Stiglitz and Rosengard, 2015: 93]. Externalities arise when individual actors take an action that generates spillover effects into the public domain for which they do not bear the full costs or from which they do not reap the full benefits. Thus, externalities are important inputs into PGs and are explicitly referred to here only when the context warrants it.

[6]Public good provision can be viewed as composed of two different though closely interrelated policy processes: (i) the political process of deciding which goods to provide, which policy paths to follow, and how to share the costs and benefits; and (ii) the implementation or production process of mobilizing and assembling all requisite inputs, including the required financial resources.

[7]For early contributions to the state-centric approach to PG provision, see, among others, Musgrave [1973, 1989] and Samuelson [1954, 1955]. Notable early contributions on private and voluntary PG provision include Buchanan [1954, 1965]. On these contrasting visions, see also Buchanan and Musgrave [1999].

[8]The concept of "social aggregation technology" was introduced by Hirshleifer [1983]. For its further refinement and application, see Cornes and Sandler 1996 [1986] and Sandler [1998].

also examine the good to identify its various building blocks and any external forces that may impact its availability. Public goods are now accepted as being public in consumption and provision.

2.2. *Key features of global public goods*

As mentioned, many scholars assume that the difference between NPGs and GPGs lies mainly in their different geographic and temporal reach. The benefits and costs of GPGs are viewed as spanning the world and, in many instances, several generations as well. This explains why the terms "inter-national PG", "trans-national PG", and GPG are often used interchangeably. However, the implications of a PG's worldwide and multi-generational reach reveal that the difference between NPGs and GPGs is more profound and that the terms "inter-national" and "trans-national" may be inadequate, and may even conceal the essential differences.

To elaborate, a good being global public in consumption implies that it confronts different countries and population groups with the sameness and indivisibility of supply, ignoring any differences in living conditions and preferences, which also tend to be wider in a global than in a national context. Moreover, in many instances, GPGs penetrate countries without asking for permission. They are not just waiting at the inter-national level and spanning across countries, patiently waiting like Weisbrodian option-value goods to be accessed, should potential users wish to do so.[9] Rather, they just transgress national borders, affecting, for better or worse, the welfare and well-being of countries or particular population groups within them. Thus, GPGs may be perceived as running counter to one of the basic principles on which the current world order is based: non-interference by outside forces in internal national affairs.

Moreover, GPGs run counter to a second fundamental world-order principle: national policymaking sovereignty. Like their

[9] Examples of Weisbrodian option-value goods are public parks and hospitals. While people may be uncertain about when or whether at all they may use these goods, should they ever wish to do so, the goods are there for them. For a more detailed discussion of this type of goods, see Holtmann [1999] and Weisbrod [1964].

national counterparts, GPGs are public in provision, yet publicness in provision means that, if a country wants to alter the supply level or the design and shape of a GPG, unilateral action or forming a small group of like-minded states may not suffice. The country may be compelled to seek the cooperation of others; GPGs that follow a summation process confront states with a compulsion to engage in effective international cooperation.

Moreover, a good's global publicness in consumption often affects not only countries but also areas beyond national jurisdiction, including many global natural commons such as the high seas and the atmosphere. The systemic integrity requirements of these commons may demand more in terms of corrective action than all states together are willing to do for them.[10]

Further, some GPGs are stock variables, whose availability changes slowly over long periods of time so that their present supply level depends on what happened in the past, and their future supply may be determined by policy actions taken today [Nordhaus, 2006].[11]

Thus, as Box 2 illustrates, GPGs have a horizontal (geographic and temporal) dimension and a vertical (border-transgressing and interdependence-generating) dimension. To capture both, and as many global challenges have national and international or transnational dimensions (and often inter-temporal ones), it seems appropriate to refer to them as global public goods (GPGs) and to use "global" to denote the goods' multi-dimensional and (by implication) complex nature: they span the world and, possibly, also several

[10] "Systemic integrity requirements" refers to the conditions that need to be maintained or achieved in order for a system (natural or human-made) to function in an intact and undiminished way. See also Costanza and Mageau [1999] and Lu *et al.* [2015].

[11] Important is that stock variables like climate change raise not only intergenerational equity issues but also inter-country issues, as we can see from the discussions on the principle of common but differentiated responsibilities and respective capacities (CBDRRC) and the linked issue of who ought to take leadership in addressing this challenge. See Brunnée and Streck [2013], CESR and TWN [2015] and Pauw *et al.* [2014].

generations, cutting across several levers of governance and are thus most likely of concern to a number of diverse actor groups.

Box 2: Defining Global Public Goods

Global public goods (GPGs) share with other public goods (PGs) the key property of publicness in consumption: being fully or partially non-rival and non-excludable. What distinguishes them from other PGs is the reach of their publicness in consumption, which: (i) spans several geographic regions or even the globe as a whole; and may also (ii) penetrate into countries, areas beyond national jurisdictions, or both, with variable levels of impact; and (iii) be of long-term duration, affecting, for better or worse, several generations. Thus, while criterion (i) is the prerequisite for a good to be defined as a GPG, the publicness in consumption of GPGs could potentially comprise three dimensions:

- A spatial dimension: being of worldwide span;
- An impact dimension: affecting countries and areas beyond national jurisdiction; and
- A temporal dimension: having long-term effects.

In most cases, global publicness in consumption along any of these three dimensions will not be an innate property of the good but reflect a policy choice or the lack thereof.

In addition to being public in consumption, many GPGs, like other PGs, are also public in provision: their provision involves a large number of actors and compels countries to seek the cooperation of others.

Source: Kaul *et al.* [2016], Introduction.

That said, however, what does the fact that climate change is such a GPG-type issue tell us about climate finance?

The next section will examine this question.

3. Climate Finance Seen through the Lens of Global Public Goods

The fact that both climate change (the problem) and climate change mitigation and adaptation (the policy outcome to be achieved) possess GPG characteristics has important implications for our understanding of and policy practice for climate finance. This is evident from the following analysis, which shows concerning four aspects of climate finance that employing the GPG lens makes a difference. The four aspects pertain to (1) climate finance as distinct from development finance; (2) the risk of market and state failure in providing climate finance; (3) the importance of intra-generational fairness as a means of strengthening actors' willingness to "get the price of international cooperation right"; and (4) the need for a more genuinely global, integrated, and comprehensive approach to the analysis and management of the provision and financing of climate change mitigation and adaptation.

3.1. *Climate finance as distinct from development finance*

In their classic book, *Public Finance in Theory and Practice*, R. A. Musgrave and P. B. Musgrave [1989 [1973]] distinguish between an efficiency branch and an equity branch of national public finance. While the former is concerned with the efficient allocation of resources to the provision of NPGs, including the management of externalities, the latter concerns how to support society in realizing what it considers a fair or just state of distribution. In line with this distinction, one could say that IPCF constitutes the international component of the efficiency branch of global public finance and that external official development finance (EODF) is the international extension of the equity branch of national public finance.[12]

[12]The public components of IPCF and EODF are flows that have budgetary implications for the contributing country. They may be provided through various instruments and institutions. For listings of the finance streams counted as "public" and "private" in the field of climate finance, see, among others,

One could say this because climate change and its mitigation are global-public in consumption and production, concerning virtually all countries and all people. When actors come together to discuss and agree on how to address this issue, the good to be produced ought to be at the center of their debates, including questions such as which strategy to choose to get to the desired policy outcome and how to share the costs and benefits of taking corrective action among the concerned parties. In contrast, when actors come together to discuss development, their focus would need to be on a particular country or group of countries like, for example, least developed countries, considering that country conditions vary and, as past experience has shown, effective development assistance requires taking national circumstances into account.[13]

Table 1 displays the significant differences between IPCF and EODF, which, if ignored, could lead to misallocations and negatively impact both branches.

Undoubtedly, synergy effects, too, appear between the two branches. As also shown in Table 1, effective development depends on effective climate change mitigation and adaptation, which, in turn, depend on effective development. Precisely due to the benefits of these synergies, however, it is important not to promote one goal at the expense of the other, as happens at present.[14]

Buchner *et al.* [2015], OECD and CPI [2015], and UNFCCC/SCF [2014]. For the categorization of development finance according to these criteria, see also https://data.oecd.org/drf/other-official-flows-oof.htm/.

[13] A country-specific approach to development assistance does not preclude policy dialogue between the national authorities and their external partners about what are the most promising approaches to fostering the country's national development. It just implies avoiding over-standardized policy prescriptions. Yet, when actors demand or offer to supply inputs to a GPG like climate change mitigation and adaptation, it would be important to be specific and, even, prescriptive about the quantitative and qualitative dimensions of the input in question in order to facilitate a smooth functioning of exchanges such as the trading of certified emission-reduction credits. We will revert to this issue below, in point 3.3.

[14] It is perhaps useful in this context briefly to discuss the increasing use of EODF for purposes that are not directly related to the conventional goal of laying a basic development foundation into countries to enable them to promote

Table 1. Climate Finance as Distinct from Development Finance.

Dimension	International public finance (IPCF)	Official development support finance (ODSF)
Main rationale	Efficiency considerations, motivated by self-interest or mixed considerations (self-interest and other-regarding concerns)	Concern about others, notably developing countries and poverty reduction in these countries
Main focus of the intervention	The GPG (global public good) to be produced	A particular developing country or developing countries more generally
Cooperating parties	Interested state and nonstate actors from all or several parts of the world	Rich and poor countries, plus perhaps their respective partners
Nature of the intervention	Multiple policy approaches, depending on the input to be provided and the implementation strategy chosen: ○ joint, collective international-level action (e.g., global norm setting or creating a co-sponsored operational or finance mechanism) ○ quid-pro-quo exchange of national or regional policy reform commitments ○ market-type exchange involving national delivery of an input against payment of a "price" ○ compensatory payment for adjustment measures and other climate-related loss and damage incurred by poorer countries	Transfers of financial and non-financial resources from richer to poorer countries
Main intended beneficiaries	One's self and, in the case of mixed-motive or altruistic actors, possibly also future generations and the Earth as a whole	Developing countries, possibly also their poor people

(*Continued*)

Table 1. (*Continued*)

Dimension	International public finance (IPCF)	Official development support finance (ODSF)
Determinants of an effective intervention	Fairness and justice in international negotiations and expectation of intra- and inter-generational outcome fairness plus willingness and capacity (i.e., development) on the part of all concerned countries	Development-compatible provision of relevant GPGs, including climate change mitigation and adaptation

Source: Author.

For example, OECD/DAC makes special efforts to document the increasing amount of EODF from its member states that flows into supporting climate change mitigation and adaptation projects and programs. In 2013–2014, bilateral climate-related ODA commitments reached on average per year USD 25 billion or 18% of total bilateral ODA, up from 4% in 2003–2004 [OECD/DAC 2015]. This is justified by the argument that, at the country level, it is difficult to distinguish

their own national development and contribute to various regional and global policy goals. Besides the allocations to climate-related purposes, these non-conventional uses of EODF include, among others: fighting poverty in middle-income countries (MICs); undertaking projects linked to military interventions and peace-keeping; and paying for the cost of asylum-seekers from developing countries in industrial countries. These expenditures are, in many instances, aimed at addressing cross-border spillover effects of human neglect, war and conflict. They are thus GPG-related and should be charged to the budget of the concerned ministry/department of the "donor" countries, like, for example, the ministry/department of defense, environment, or interior and home security affairs. See, on these points: the OECD/DAC High Level Meeting Communiqué of 19 February 2016, which endorses the use of official development assistance (ODA) for these various non-conventional purposes (available at http://www.oecd.org/dac/DAC-HLM-Communique-2016.pdf/); for voices arguing for ODA to continue focusing on the least-developed countries and cooperation with MICs to be funded from alternative sources, see, among others, Glennie [2011] and Tomasi [2015], as well as the categorization of "development support" suggested by Lin and Wang [2015].

between development assistance and climate finance. True, at the country level, the IPCF and EODF streams often come together and are, even, pooled with domestic public, private and other types of finance.

Nevertheless, in order to ensure that each funding stream flows to where it can best be used and to demonstrate that IPCF consists of new and additional resources, as repeatedly stipulated in international agreements, it would be desirable and possible clearly to distinguish between them — upstream. When preparing their national budget, all donor countries could allocate IPCF to one budget line and EODF to another line. Development assistance could be distributed, via bilateral, multilateral and other channels, to qualifying countries according to defined allocation criteria; and it would be the recipient developing country, who would, as called for in Agenda 2030,[15] report on the progress it has achieved in national capacity building and advancing towards the Agenda goals, including an overview of related expenditures, as and if appropriate, broken down by funding source. In contrast, IPCF should be programmed on an issue basis, for onward channeling to actors (countries, firms, civil society organizations, individual experts), who appear to be the relatively best providers for a particular input to the good to be produced.[16]

To avoid inefficient intertwining of climate finance and development finance, it could also be useful to establish an OECD/DAC sister organization to which governments would report GPG-related expenditures, including climate finance allocations.

Initially, in the early 1990s, when the growing importance of GPGs became increasingly apparent, it was perhaps a pragmatic move to address urgent global challenges by tapping ODA resources, but why does this practice continue? Why not recognize climate finance as distinct from development assistance?

---

[15]See on Agenda 2030 again note 2 above.

[16]For a more detailed discussion on these resource allocation processes, see Kaul *et al.* [2015].

The next section provides one answer to this question.

3.2. *The risk of market and state failure in ensuring adequate climate financing*

Standard economics assumes individual actors to be rational and selfish, primarily interested in improving their own material welfare. As such, they are predicted to free-ride on PGs, a behavior that is seen to cause markets to fail and potentially call for the state to step in with corrective action through its coercive powers of regulation and taxation. However, the institution of the state has no international equivalent. In fact, at the international level, states are also individual actors, and their representatives in international negotiations are likely to pursue national interests.

Many scholars therefore assume that, just like any other individual actor, states are rational, primarily self-interested and, in the presence of GPGs, tempted to free-ride, notably when viewing the issue under negotiation as not a national priority. Put differently, GPGs are seen as facing the risk of dual (i.e., market and state) failure.

More recent research has shown that this risk, like the free-riding risk generally, may be overstated and that states are likely to be mixed-motive actors, possibly more so than individuals, because their policy positions tend to be a mixture of the interests of their various national constituencies. Moreover, states' reluctance to cooperate can reflect not free-riding but a rejection of an agreement proposal perceived as unattractive.

For a variety of reasons, then, dual-actor failure in the presence of GPGs may occur, especially concerning finance issues such as IPCF. A government may view national revenue as "theirs", a good that they can mobilize, appropriate, and control in terms of the uses made of it. National public revenue is one of the power bases of states, but this has weakened in recent decades, as various tax rates have lowered, and states have relied on debt financing of their budgets (see, Ostry *et al.*, 2015).

Could these factors explain why, for example, many donor countries promise to move towards the ODA target of 0.7% but repeatedly fail to do so and, in addition, even tug climate finance into the ODA envelope?[17] Or, why they are keen to mobilize private finance for climate purposes without precisely clarifying what the comparative advantages of different actor groups and funding sources in this field are? If discussed at all, the role of public finance that is mainly mentioned is that of catalyzing private finance. Yet, especially in policy areas such as climate change, which call for transformative change, states — individually and collectively — would need to play a proactive, entrepreneurial role that goes far beyond leveraging private finance, however important, perhaps even indispensable this is.[18]

Enhancing the efficiency and effectiveness of climate finance might thus require further empirical and theoretical research aimed at developing an economic theory that explains under which conditions particular state and non-state actor groups contribute or fail to contribute to GPGs and what their comparative advantages are in their contribution to various types of financial and non-financial inputs. The experiences such as those gained with public-private partnerships and the analyses of the current flows of climate finance, such as the limited amounts of resources so far allocated to adaptation purposes, suggest that the funders and their interests may matter: Neither private finance nor voluntary finance are perfect substitutes for public finance. It would thus be important systematically

[17] At the writing of this chapter, only five out of 29 OECD/DAC member states have met the ODA target of 0.7% of their national income on a continuing basis. See http://www.oecd.org/dac/stats/ODA-history-of-the-0-7-target.pdf/. Yet, about 18% of total bilateral ODA (amounting to an annual average of USD 25 billion) were already flowing to climate-related initiatives in 2013–2014. See http://www.oecd.org/dac/environment-development/Climate-related-dev-finance-ENG.pdf/. This amount is relatively modest compared to the total amount of climate finance, which Buchner *et al.* [2015: 1] estimate to amount to around USD 390 billion in 2014. Therefore, its strategic use would be all the more important.

[18] On the role of the entrepreneurial state, see Mazzucato [2015].

to clarify the specificity of the roles played by different actor groups in climate financing.

In this context, it would also be important to reexamine the nature of international cooperation, notably multilateral (i.e., inter-state) cooperation, and to explore whether it is just the joint, collective endeavor (as we tend to view it) or whether it might be more appropriately viewed as a market transaction — an exchange between a country (group) A and a country (group) B, in which A's commitment to reform is being traded against matching policy commitments and/or financial transfers to be provided by B. This question poses itself since, with the rising trend towards multipolarity, conventional power politics is increasingly giving way to "smart power" politics. In light of this trend, it seems reasonable to conjecture that the top-down policy pressure that often enforced international agreements in the past has been replaced by exchange-type interactions between states.

Therefore, to what extent is international cooperation functioning as an international political market, and how efficient is this market?

This brings us to the next issue.

3.3. *The importance of intra-generational fairness*

As Stern [2015] points out, many researchers, policymakers, and experts in the field of climate change emphasize concerns about intergenerational equity. Yet, so Stern, given that scientists warn us that corrective action is urgently required if we are to avoid crossing the 2°C limit of global warming and, preferably, even stay below this threshold, strong emphasis must also, if not primarily, be placed on intra-generational equity within and across countries as a means of fostering the requisite willingness to contribute among all the concerned parties. This is especially urgent where actors' preferences vary across countries and population groups (as is undoubtedly the case concerning many facets of climate change mitigation) and where decisive corrective action is required now (see also Rübbelke, 2011).

In these cases, implementation strategies would need to aim at producing mixed-benefit results — clear near-term net benefits (e.g., clean-energy security for all), which, as co-benefits, contribute to the "climate change mitigation" stock variable.

We are living in a multi-polar world, and fairness and justice in international cooperation can no longer mean just offering any type of side-payment, a "carrot", as game theorists say. Nor can it mean that the conventional powers wield "sticks" (another game theory term) such as trade sanctions that threaten to make non-compliance with a proposed agreement more expensive than compliance. Rather, current global power balances call for more systematic, transparent, and disaggregated cost–benefit assessments and, importantly, bargains that are perceived as worthwhile and fair — that is get the "price" right.

As Box 3 shows, each of the main cooperation strands — joint, collective action, and policy-commitment exchanges — involve several types of financial transaction that need to be calculated transparently and according to established fairness principles. This is especially important for exchanges where the suppliers of inputs are often developing countries and developed countries figure on the demand side as financiers. For bargains or sales to be concluded, it is important that the suppliers see that a reasonable amount of fairness has been factored into the prices being offered by the countries on the demand side of the bargain. Similarly, the "financiers" must have clear output indicators so that they can without too much effort or cost determine whether the good they financed is being delivered and meets all agreed-upon quantitative and qualitative criteria.

Thus, fairness and enhanced contracting may have to move in tandem.

Box 3: Possible Uses of International Public Climate Finance

International cooperation among states consists of three main sets of inter-action that tend to be supported by international public climate finance (IPCF): (1) national-level policy reform

Box 3: (*Continued*)

commitments by one country or set of countries in exchange for IPCF transfer commitments by others; (2) compensatory arrangements for domestic loss and damage resulting from global spill-ins; and (3) joint efforts of countries in support of the production of intermediate GPG-type goods that require pooled IPCF financing. Each of these cooperation strands requires different financial mechanisms and characteristics, according to which the concerned parties may consider them as being in line with intra-general fairness.

1. *Exchange-type transactions*
This type of transaction occurs when states meet each other on opposite sides of the international political market and seek to match the demand for and supply of domestic reform steps that would help generate inputs to a goal such as climate change mitigation and adaptation. The financing from the demand to the supply side could, for example, include:

- *Incentive payments* — provided by developed countries to developing countries to encourage the latter to undertake mitigation-related domestic policy reform and capacity-building initiatives that might otherwise not happen to the extent or at the speed that the former consider desirable from a global perspective; and
- *Adequate pricing of climate-related goods and services* — which can, in turn, take two forms:
  - Payments to a particular developing country for the generation of diffuse positive climate-related externalities (e.g., by preserving or expanding tropical-forest areas in its jurisdiction), provided on demand from the international community or parts thereof in excess of what the country would have done had it been motivated solely by its own national interests; these are also referred to as "reimbursements of incremental costs"; and

Box 3: (*Continued*)

- Payments for climate-related goods and services of a private-good nature (e.g., certified emission reductions) provided by a developing country on demand from a developed country.

2. *Compensatory finance*
This category includes payments by developed countries to developing countries to reimburse the latter for costs associated with adaption to climate change and other climate-related loss and damage.
3. *Collective production of intermediate regional or global public goods*
In this case, all interested countries would engage in the production of regional public goods (RPGs) and GPGs such as early warning and weather forecasting systems, research and development as well as deployment of knowledge and technology for facilitating the tackling of climate change-related and associated goals such as energy security for all, or weather and crop insurance. The financial arrangements that might support such joint or collective endeavors could include:

- *Cost-sharing arrangements* — for which the cooperating parties would need to decide the most appropriate principle to follow: ability-to-pay, beneficiary/user to pay, or any other principle;
- *Pooled subsidy payments* — from richer to poorer countries to enable the latter to establish national capacity and measures to access these goods; and
- *Pooled subsidy payments* — from richer countries to private actors (e.g., insurance companies) to encourage them to offer differentiated prices so that poorer countries could also afford their products.

In respect to each main type of transaction and each payment arrangement, it would be important to establish clear and transparent methods of calculating costs and benefits, as well as

Box 3: (*Continued*)

simple criteria and indicators that would allow policymakers and experts to determine whether a proposed transaction qualifies as a fair and mutually beneficial bargain. While precise cost–benefit calculations may sometimes be difficult, it should nevertheless be possible for the concerned parties to agree on payment and product-delivery arrangements that all perceive as "vaguely right" and hence acceptable, so that, in type 1 exchanges, bargains would be concluded; type 2 arrangements might contribute to a sense of fairness among the concerned parties; and, for type 3 collective endeavors, the desired intermediate RPGs and GPGs would emerge as promptly and adequately provided as possible.

http://unfccc.int/files/essential_background/convention/application/pdf/english_paris_agreement.pdf

Source: Author.

Interestingly, as can be seen from the Agenda 2030 and the Paris Agreement,[19] we appear to have progressed further on indicators and result measurement than on fair pricing, considering that the very definition of "climate finance" has remained unsettled, incremental costs are often viewed as too complicated to determine, and considerable amounts of IPCF are being charged to ODA accounts. Expected results are usually being defined with great precision. When it comes to finance, however, matters are left open-ended and vague. For example, Article 9 of the Paris Agreement just states that developed countries shall provide financial resources to developing countries in continuation of their obligations and encourages other parties to provide support voluntarily.

Does this vagueness indicate that international political markets are suffering from conditions similar to those that make economic

---

[19]See, note 1 for the reference to the agreements again.

markets fail, such as power concentration and information asymmetries, and is the growing attention being paid to the formation of "clubs of the willing" introducing further distortions and preventing the enhanced efficiency in these international political markets that could come from pricing fairly and transparently and in a way that respects agreed-upon principles such as the CBDRRC principle?

4. Integrated Provision Path Analysis and Management

Most of us might take for granted that a producer of a simple private good such as a pencil or paper clip had undertaken a production path analysis to determine the required inputs, from where best to procure them, and when and how to feed them in the production process. However, though innumerable studies have examined many relevant scientific, technological, economic, socio-cultural, and ethical facets of GPGs, we still lack for these goods provision path sketches indicating, in broad-brush strokes, the major types of public/private and national/international inputs that might be required, their proper sequence, and the appropriate speed at which to move towards producing them.[20]

Such a provision path sketch could serve several purposes. For example, it would clarify which actor groups might be best suited to provide which inputs, thus indicating the desirable balance between

[20]The report of Working Group III to the Fifth Assessment Report of the Intergovernmental Panel on Climate Change [IPCC, 2014] is a case in point. It presents an impressive overview of the various facets of climate change mitigation but discusses them one-by-one without taking the next step of drawing up a provision path map. Other studies (e.g., Edenhofer *et al.* 2015) refer to the importance of creating links between various building blocks of the good but do not yet indicate what linking could entail.

However, most of the more comprehensive provision path analyses take the form of graphs, as in Kaul and Conceição [2006] and Kaul *et al.* [2016], for example. Nevertheless, these studies, as well as many of those that provide micro-analyses of isolated facets of GPG provision, provide a rich source that could be tapped to prepare more comprehensive provision path analyses, though some of the findings might need reexamination to ascertain whether they are applicable to the GPG context.

public and private finance. In this case, it would not be the availability or supply of certain types of money that would drive implementation. Rather, money would follow, and serve, the identified policy paths.

Of course, preparing such sketches requires clear goal-setting. For example, should climate change mitigation be achieved primarily through scarcity management such as the establishment of quantitative global and national pollution-reduction targets or by aiming at goals like, for example, "clean-energy security for all" or "equal access for all to opportunities for sustainable growth and development"? Obviously, clarifying the goal is the starting point of any provision path analysis, and the strategy chosen and financed will matter to intra-generational equity and the mobilization of the requisite willingness to cooperate and pay among all concerned parties. Is it, then, because of the still-lacking process and output fairness that, despite the warnings that time is running out, climate governance is still highly fractured and haphazard?[21]

Provision path sketches would also allow a systematic exploration of subsidiarity issues and the identification of opportunities for economies of scale and scope and for the more efficient use of non-rival goods such as green technology.[22]

However, we not only lack provision path sketches for GPGs but also agents or agencies who could function as central nodes of what would in many cases be undoubtedly highly complex (i.e., multi-actor, multi-sector, multi-level) provision networks. Such an agent or agency would be required to change provision path analyses from being mere studies into a living document — an instrument for introducing a sense of direction into GPG production processes. In the field of climate change, such a facilitation mechanism exists for

[21]The notion of "process fairness" refers to parties having an effective say in matters that concern them; and output fairness refers to the parties, who are involved in a cooperative effort, to consider the resulting policy output as fair.

[22]See Stiglitz [2014] on the management of global knowledge stock, and Maskus and Reichman [2005] on technology transfer and the global intellectual property rights regime.

the political dimension (i.e., international negotiations) but not for the production dimension. This governance void needs to be filled.

Recognizing GPGs, preparing GPG provision path sketches, appointing GPG network facilitators, and introducing "global" to national and international governance entities as a new organizational criterion could go a long way towards facilitating a more integrated and efficient implementation process, as well as an enhanced financing of these goods. While not involving any major change, it would introduce an important paradigm shift in research and policymaking by replacing actors such as governments, states, firms, civil society, or individual households at the center of the analysis with the good, the policy outcome — the GPG to be achieved. This shift in focus would facilitate the often sorely lacking integration of the myriad dimensions that typically characterize complex issues like climate change.

5. Conclusion

This paper has explored whether there is added value in examining climate change mitigation and adaptation through the analytical lens of GPGs. The discussion indicates that this question can be answered in the affirmative. We noticed, by looking through the analytical lens of GPGs, that IPCF and, thereby, overall climate finance seems to suffer from theoretical and institutional lock-in — a continued reliance on theories and policy practices that fit neither the GPG nature of climate change nor current policymaking realities. The discussion points to a rich agenda for further research and debate on how to improve the organization and governance of climate finance.

The analysis also shows, however, that the constraints facing climate finance are, by and large, not climate-specific but GPG-specific, affecting GPG-type challenges across the board. Yet, a general theory of GPG provision and finance has not yet been developed.[23]

[23]Beginnings of such a theory can be found in Atkinson [2006], Kaul and Conceição [2006], and Sandmo [2007, 2015].

Therefore, it would seem useful to pursue a two-pronged approach to overcoming the current lock-in problems: developing the incipient effort made in this chapter to apply the GPG lens to climate issues while concurrently building a general theory of GPG provision and finance, which could offer robust policy advice on the cross-cutting institutional reforms that might be required. Such a two-pronged strategy suggests itself, because it seems unlikely that policymakers and other actors would support reform steps such as fairness or a mutually considerate and respectful exercise of national policymaking sovereignty in one particular issue area —say, climate— and not in other areas such as global health and finance, or managing the global natural commons, outer space, and cyber security.

In fact, developing such a general theory of GPG provision could go a long way towards avoiding the risk of considering all financial resources, irrespective of their source, as mutually substitutable. It would allow us to determine more clearly the specificity of the roles that different actors are best poised to assume in the provision process of a particular good. To be clear about that is especially important in today's multi-actor, in which international cooperation endeavors such as climate change mitigation and adaptation can benefit from public, private and voluntary resource flows but in which the role of public finance, notably that of IPCF, is especially important in order to ensure that the ultimate policy outcome is not only global-public in consumption and provision but also generally perceived as fair and just.

Moreover, the time for developing a general GPG theory appears to be ripe. In most global-challenge areas, innumerable studies offer highly specialized micro-analyses of a particular facet of a global challenge. Many of their insights will undoubtedly remain valid and provide a foundation on which to build further research. As many of their findings pertain only to either the national or international level, however, they may require reexamination from a GPG perspective, notably from the perspective of how to provide and finance GPGs in today's increasingly interdependent, multi-polar and multi-actor world.

References

Atkinson, T. 2006. Global Public Finance. *Oxonomics* 1, 2–4.

Brunnée, J. and C. Streck. 2013. The UNFCCC as a Negotiation Forum: Towards Common but More Differentiated Responsibilities. *Climate Policy* 13 (5), 589–607.

Buchanan, J. M. 1954. Social Choice, Democracy, and Free Markets. *Journal of Political Economy* 62, 114–123.

Buchanan, J. M. 1965. An Economic Theory of Clubs. *Economica* 32, 1–14.

Buchanan, J. M. and R. A. Musgrave. 1999. *Public Finance and Public Choice.* Cambridge, MA: MIT Press.

Buchner, B., C. Trabacchi, F. Mazza, D. Abramskiehn and D. Wang. 2015. *Global Landscape of Climate Finance 2015.* Climate Policy Institute. Venice, Italy.

Center for Economic and Social Rights (CESR) and Third World Network (TWN). 2015. *Universal Rights, Differentiated Responsibilities: Safeguarding Human Rights beyond Borders to Achieve Sustainable Development Goals.* Human Rights Policy Briefing. CESR and TWN. Brooklyn, NY, and Penang, Malysia.

Cornes, R. and T. Sandler. 1996 [1986]. *The Theory of Externalities, Public Goods and Club Goods.* Cambridge, UK and New York, NY: Cambridge University Press.

Costanza, R. and M. Mageau. 1999. What Is a Healthy Ecosystem? *Aquatic Ecology* 33,105–115.

Desai, M. 2003. *Public Goods: A Historical Perspective.* In Providing Global Public Goods: Managing Globalization, I. Kaul, P. Conceição, K. Le Goulven and R. U. Mendoza (eds.), New York, NY: Oxford University Press, 63–77.

Edenhofer, O., M. Jakob, F. Creutzig, C. Flachsland, S. Fuss, M. Kowarsch, K. Lessmann, L. Mattauch, J. Siegmeier and J. C. Steckel. 2015. Closing the Emission Price Gap. *Global Environmental Change* 31, 132–143.

Glennie, J. 2011. *The Role of Aid to Middle-Income Countries: A Contribution to Evolving EU Development Policy.* Working Paper 331. Overseas Development Institute. London, UK.

Hirshleifer, J. 1983. From Weakest-Link to Best-Shot: The Voluntary Provision of Public Goods. *Public Choice* 41 (3), 371–386.

Holtmann, A. G. 1999. Samulsonian and Weisbrodian Public Goods. *Public Finance Review* 27 (4), 387–395.

Intergovernmental Panel on Climate Change (IPCC). 2014. *Climate Change 2014: Mitigation of Climate Change.* Contribution of Working Group III to the Fifth Assessment Report of the Intergovernmental Panel on Climate Change. Cambridge, UK and New York, NY: Cambridge University Press. Kaul, I., D. Blondin and N. Nahtigal. 2015. *Country Allocations Versus Issue Allocations: The Case of Climate Finance.* In Financing Sustainable Development: Addressing Vulnerabilities, M. Boussichas and P. Guillaumont (eds.), Paris, France: Economica, 353–370.

Kaul, I., D. Blondin and N. Nahtigal. 2016. *Global Public Goods.* The International Library of Critical Writings in Economics. Cheltenham, UK: Edward Elgar.

Kaul, I. and P. Conceição, eds. 2006. *The New Public Finance: Responding to Global Challenges.* New York, NY: Oxford University Press.

Kaul, I., P. Conceição, K. Le Goulven and R. U. Mendoza, eds. 2003. *Providing Global Public Goods: Managing Globalization.* New York, NY: Oxford University Press.

Kaul, I. and R. U. Mendoza. 2003. *Advancing the Concept of Public Goods.* In Providing Global Public Goods: Managing Globalization, I. Kaul, P. Conceição, K. Le Goulven and R. U. Mendoza (eds.), New York, NY: Oxford University Press, 78–111.

Lin, J. Y. and Y. Wang. 2015. *China's Contribution to Development Cooperation: Ideas, Opportunities and Financing.* In Financing Sustainable Development: Addressing Vulnerabilities, M. Boussichas and P. Guillaumont (eds.), Paris, France: Economica, 73–89.

Lu, Y., R. Wang, Y. Zhang, H. Su, P. Wang, A. Jenkins, R. C. Ferrier, M. Bailey and G. Squire. 2015. Ecosystem Health towards Sustainability. *Ecosystem Health and Sustainability* 1 (1), Article 2.

Maskus, K. E. and J. H. Reichman. 2005. *International Public Goods and Transfer of Technology under a Globalized Intellectual Property Regime.* Cambridge, UK: Cambridge University Press.

Mazzucato, Mariana. 2015.*The Entrepreneurial State: Debunking Public vs. Private Sector Myths.* New York, NY: Public Affairs.

Musgrave, R. A. and P. B. Musgrave. 1989. *Public Finance in Theory and Practice.* 5th edn. New York, NY: McGraw-Hill.

Nordhaus, W. D. 2006. Paul Samuelson and Global Public Goods: A Commemorative Essay for Paul Samuelson. In Samuelsonian Economics and the Twenty First Century, M. Szenberg, L. Ramrattan and A. A. Gottesman (eds.), New York, NY: Oxford University Press, 88–98.

Organisation for Economic Co-operation and Development (OECD). 2015. *Climate Finance in 2013–14 and the USD 100 Billion Goal.* A report by the OECD in collaboration with Climate Policy Initiative (CPI). OECD. Paris, France. At: http://climatepolicyinitiative.org/wp-content/uploads/2015/10/OECD-2015-Climate-Finance-Report-with-CPI-LR.pdf/.

Organisation for Economic Co-operation and Development/Development Cooperation Directorate (OECD/DCD). 2015. *Climate-Related Development Finance in 2013–14.* OECD. Paris, France. At: https://www.oecd.org/dac/environment-development/Climate-related-dev-finance-ENG.pdf/.

Ostry, J. D., A. R. Ghosh and R. A. Espinoza. 2015. *When Should Public Debt Be Reduced?* Staff Noes No. 15/10. International Monetary Fund. Washington, D.C.

Pauw, P., S. Bauer, C. Richerzhagen, C. Brandi and H. Schmole. 2014. *Different Perspectives on Differentiated Responsibilities: A State-of-the-Art Review of the Notion of Common but Differentiated Responsibilities in International*

Negotiations. DIE Discussion Paper 6/2014. German Development Institute. Bonn, Germany.

Rübbelke, D. T. G. 2011. International Support of Climate Change Policies in Developing Countries: Strategic, Moral and Fairness Aspects. *Ecological Economics* 70, 1470–1480.

Samuelson, P. A. 1954. The Pure Theory of Public Expenditure. *The Review of Economics and Statistics* 36 (4), 387–389.

Samuelson, P. A. 1955. Diagrammatic Exposition of a Theory of Public Expenditure. *The Review of Economics and Statistics* 37 (4), 350–356.

Sandler, T. 1998. Global and Regional Public Goods: A Prognosis for Collective Action. *Fiscal Studies* 19 (3), 233–244.

Sandmo, A. 2007. *The Welfare Economics of Global Public Goods.* NHH Dept. of Economics DP 35/2007. Norwegian School of Economics and Business Administration. Bergen, Norway.

Sandmo, A. 2015. *The Public Economics of Climate Change.* NHH Dept. of Economics DP 27/2015. Norwegian School of Economics. Bergen, Norway.

Stern, N. 2007. *The Economics of Climate Change; The Stern Review.* Cambridge, UK: Cambridge University Press.

Stern, N. 2015. *Why Are We Waiting? The Logic, Urgency, and Promise of Tackling Climate Change.* Cambridge, MA: MIT Press.

Stiglitz, J. E. 2014. *Intellectual Property Rights, the Pool of Knowledge, and Innovation.* NBER Working Paper 20014. National Bureau of Economic Research. Cambridge, MA.

Stiglitz, J. E. and J. K. Rosengard. 2015. *Economics of the Public Sector.* 4th edn. New York, NY and London, UK: W.W. Norton.

Tomasi, S. 2015. *Modernising and Revitalising the Measurement of Development Finance.* In Financing Sustainable Development: Addressing Vulnerabilities, M. Boussichas and P. Guillaumont (eds.), Paris, France: Economica, 59–72.

United Nations Framework Convention on Climate Change/Standing Committee on Climate Finance (UNFCCC/SCF). 2014. *2014 Biennial Assessment and Overview of Climate Finance Flows Report.* Bonn, Germany: UNFCCC Secretariat.

Weisbrod, B. A. 1964. Collective-Consumption Services of Individual Consumption Goods. *Quarterly Journal of Economics* 78 (3), 471–477.

Chapter 8

International Adaptation Funding and the Donor's Welfare Maximization

Oliver Schenker

Frankfurt School of Finance & Management,
Sonnemannstraße 9-11,
60314 Frankfurt am Main, Germany
o.schenker@fs.de

Gunter Stephan

Universität Bern, Department of Economics &
Oeschger Centre of Climate Change Research,
Schanzeneckstrasse 1, 3001 Bern, Switzerland
gunter.stephan@vwi.unibe.ch

Abstract

In a recent paper, Schenker and Stephan [2014] have shown that adaptation funding in the range as envisaged in international climate policy negotiations can be Pareto-improving. Not only will the funded developing world, which does not own sufficient resources for adapting optimally, profit from receiving adaptation funding. Terms-of-trade improvements dominate the transfer costs and hence lead to welfare gains in almost any developed region except North America. However, funding of adaptation has a public good character. It improves the production technology, reduces the production prices and improves terms-of-trade for all net-importers of vulnerable goods. Having a simple analytical model and a numerical model at hand, we show that despite its public good character Nash equilibria exist which include positive amounts of North-South adaptation funding. However, the resulting transfers are much smaller than the aspired amount of international climate finance.

1. Introduction

At the 2010 United Nations Framework Convention on Climate Change (UNFCCC) Conference of the Parties in Cancun (COP16), the Green Climate Fund (GCF) was formally established. This Fund, which redistributes contributions from developed to developing countries, is a cornerstone of the UNFCCC Long-Term Climate Finance work program and has two major aims: First, it should financially assist the developing countries in responding to the threat of global climate change through reducing carbon emissions on one hand and through increasing the climate resilience of their economies on the other. Second, it should give the developed world an opportunity to take responsibility for their role in causing climate change. The present atmospheric concentration of greenhouse gases is primarily an echo of the historical process of industrialization in the developed economies, and although least responsible the developing countries are the most heavily affected by climate change. Seen from this perspective, contributions to the GCF are a kind of voluntary compensation of an inequality that many developing nations regard as unfair.

Several papers discuss the issue of financially supporting less developed countries in coping with climate change. For example, Carraro *et al.* [2005] analyze the role of transfers as strategic instrument in the process of negotiating an international environmental agreement. Hoel [2001] shows that transfers can be an important measure for reducing carbon leakage. Pittel and Rübbelke [2013] demonstrate that financial transfers can help to address the perceived unfairness associated with historical emissions and thus facilitate negotiations to reach a comprehensive and stringent climate agreement, while Grasso [2010] as well as Eyckmans *et al.* [2015] examine the role of transfers when the motivation to transfer is ethical rather than strategic. Their results are quite similar to the ones Buob and Stephan [2013] have observed, when considering that issue from a pure strategic perspective without taking fairness considerations into account.

A common message of these papers is that for purely economic reasons the developed world has almost no incentive to voluntary donate. Reality seems to support this hypothesis. So far, only a few nations including Japan, the UK, France, and Germany have guaranteed a total of approximately USD 10 billion to the GCF. This is little compared to the goals of the Fund, which are much more ambitious. The UNFCCC hopes to collect a total of USD 100 billion per year by 2020 under its Long-Term Climate Finance work program where the GCF is not the only one but one of its core vehicles. Many observers believe that it will not be possible to get such an amount on an honor system alone. This immediately leads to the question: Do further incentives exist besides of fairness and ethical reasoning that can motivate developed countries to voluntary contribute to the GCF?

Economies do not only interact because of their responsibilities for climate change and through financial transfer for funding adaptation and mitigation. They are also embedded in a web of mutual trade relations. This web is becoming increasingly denser. Since 2000, the volume of international trade has increased by almost 90% [CPB, 2015]. International trade links influence the cost-benefit ratio of financial support measures through two channels. On the one hand, additional resources in the recipient country lead to additional demand of goods. Since these goods are potentially produced in international supply chains, these goods may have components sourced in the donor country. For example, Beutel [2002] shows by using a multi-region input-output model that the structural funds of the European Union (EU) which redistribute resources to poorer regions can have substantial spillovers benefiting also richer regions. On the other hand, as Schenker [2013] points out, these trade relations may cause spillovers of climate change impacts into economies far away from each other. This mechanism can be illustrated by a recent event: During the summer 2010, Russia experienced an extreme drought that caused large losses in crop yields. The Russian government reacted with export restrictions. This led to sharp price increases for staple food such as bread and pasta, in particular in the Baltic countries (Estonia, Latvia and Lithuania) that have close trade relations to Russia, where it even affected inflation rates [EC, 2013].

These interlinks provide a further rationale for adaptation funding can be derived. As Schenker and Stephan [2014] show both theoretically as well as numerically within the framework of a Computable General Equilibrium (CGE) model, funding of climate change adaptation in a range as foreseen by the GCF can be Pareto-improving if trade effects are taken explicitly into account. Not only will the funded, developing world, which does not own sufficient resources for adapting optimally, profits from receiving adaptation funding. Terms-of-trade improvements in the high and middle income donor countries will dominate the transfer costs and hence lead to net-welfare gains in almost any developed region except North America.

More precisely, Schenker and Stephan [2014] demonstrate: (1) If a donor country is net-importer of goods, which become scarcer and hence more expensive since exporting countries are not able to adapt sufficiently to climate impacts, adaptation funding is in the donors' self-interest in case that adaptation funding reduces the costs of importing vulnerable goods by more than the costs of funding; (2) Developing countries can benefit from adaptation funding if the benefits of reducing climate impacts dominate the loss in terms-of-trade in net-exporting vulnerable goods. Therefore, adaptation funding as foreseen by the GCF will be beneficial for all developing countries.

The Schenker-Stephan analysis, however, suffers from two important restrictions of generality. First, it is assumed that all recipients are budget constraint such that they cannot spend sufficient resources for adapting optimally to climate change impacts. This assumption guarantees that there is no crowding out of adaptation funding (see Buob and Stephan, 2013). On contrary, funds are completely used for investments into adaptation only. Second, all donors contribute the same share of their national gross domestic product (GDP) to adaptation funding. High income regions in Europe and Central Asia, North America, and East Asia and Pacific, respectively, provide financial assistance for adaptation in the low income regions and steadily increase their financial contributions until they hand over 0.02% of their GDP from 2030 onwards. This would imply that by 2050, USD 100 billion per year are transferred to low income regions.

The last assumption is of particular importance if the incentive problem of financial transfers is considered. Adaptation funding has features of a public good. On the one hand, it improves the productivity of vulnerable sectors, from which exporters of commodities from these sectors can profit. On the other hand, it reduces the output prices and hence improves terms-of-trade for all net-importers of vulnerable goods, independent on whether they have contributed to an adaptation fund or not. Not surprisingly therefore, the idea to internationally coordinate a grand top-down scheme largely failed. Instead, a process evolved in which individual countries make pledges on what they are willing to pay in coping with climate change. This motivates to analyze which funding rates are in the self-interest of developed countries, if (1) developing countries accept the funding unconditionally and if (2) the developed world autonomously adapts optimally but developing countries are not able to do so.

The rest of the chapter is organized as follows. Section 2 presents a simple theoretical analysis that reveals the basic principles and mechanics. With these lessons at hand, we turn to an analysis of different funding schemes by using a numerically calibrated multi-region dynamic CGE model of world trade. The model and its calibration are presented in Section 3. Section 4 discusses the results of the CGE analysis. We identify Nash equilibria where adaptation funding is part of the optimal strategy and show that the magnitude of transfers in these equilibria is much smaller than what would be needed to reach the USD 100 billion target of UNFCCC climate finance. Section 5 concludes.

2. Some Theoretical Considerations

Consider a stylized model of the world economy, where N countries trade two commodities on open international markets under perfect competition. Commodities differ with respect to their vulnerability to climate change. Some are provided by sectors, which are sensitive to climate change. Typical examples are agriculture, forestry and tourism. Other goods and services are supplied by sectors, which are almost unaffected by climate change such as telecommunication, financial services or industrial manufacturing. Countries are either donors or recipients of financial transfers. Typically donors are located in the northern hemisphere, which is a region of relatively high wealth but relatively low direct exposure to climate change. Recipient countries are usually located in the South and are part of the developing world, which is relatively vulnerable to climate change but has insufficient capacities to adapt against its impacts.

Let good v be the aggregate of commodities, which are supplied by climate-sensitive sectors and let good w be the aggregate of goods and services from sectors where climate change impacts are almost negligible. If good w is the numeraire and p the world market price of good v, then total production of region $n = 1, \dots, N$ can be characterized by the green GDP function, which corresponds to the regional GDP net of climate change impacts on vulnerable production

$$g^n(p, Q, a^n) = \max\{p\,\theta^n(Q, a^n)y_v^n + y_w^n | y_i^n = f_i^n(l_i^n), l_v^n + l_w^n \leq \bar{l}^n\}. \quad (1)$$

f_i^n is an increasing and concave production function of good $i = v, w$. l_i^n denotes the factor demand for producing good i in region r and $\bar{l}^n$ is the total factor endowment of region n, which is in-elastically supplied.

The partial derivative[1] g_p^n is the region's supply function of the vulnerable good i.e., $g_p^n = \theta^n(Q, a^n)y_v^n$ and has the properties $g_p^n > 0$, $g_{pp}^n > 0$, $g_a^n > 0$, $g_Q^n < 0$, $g_{pa}^n > 0$, $g_{pQ}^n < 0$. The last properties follow from the assumptions imposed on $\theta^n(Q, a^n)$, which is the region-specific climate impact function. It is a function of the atmospheric carbon dioxide (CO_2) concentration Q as well as the region's expenditure a^n for adapting to climate change where $\theta_Q^n(Q, a^n) < 0$ and $\theta_a^n(Q, a^n) > 0$. This means, the higher the stock Q of globally accumulated greenhouse gas emissions and the lower the adaptation expenditure a^n, the lower will be the fraction $\theta^n(Q, a^n)y_v^n$ of outputs y_v^n of good v, which is at disposal in country n.

[1]Throughout this paper subscripts indicate the partial differentiation with respect to a particular variable.

Utility maximization in region n leads to the expenditure function

$$e^n(p, \bar{U}^n) = \min\{pc_v^n + c_w^n | U^n(c_v^n, c_w^n) \geq \bar{U}^n\}. \quad (2)$$

It corresponds to the minimum expenditure, which the representative consumer in country n has to spend for achieving a given utility level $\bar{U}^n$ at price p. $c_i^n(p, U^n)$ is the conditional demand for good i in region n. The partial derivative e_p^n gives the compensated demand function of the vulnerable good and has the usual properties, i.e., $e_p^n > 0$, $e_U^n > 0$, $e_{pp}^n < 0$, $e_{pU}^n > 0$. Note that e_{pU}^n / e_U^n equals the change in demand of good v in response to a change in income.[2]

We focus on tied adaptation transfers in the sense that transfers of the numeraire good are directly used for financing adaptation. Thereby, we do not consider direct transfers from a single donor to a single recipient, but assume that a donor country $d = 1, \ldots, D$ transfers F^d units of the numeraire good to a global adaptation fund, from which a recipient country $r = 1, \ldots, R$ receives F^r units. Then the world economy equilibrium is characterized by the following system of equations:

$$e^d(p, \bar{U}^d) + a^d + F^d - g^d(p, Q, a^d) = 0, \quad d = 1, \ldots, d, \quad (3)$$

$$e^r(p, \bar{U}^r) + a^r - g^r(p, Q, a^r + F^r) = 0, \quad r = 1, \ldots, R, \quad (4)$$

$$\sum_d e_p^d(p, \bar{U}^d) + \sum_r e_p^r(p, \bar{U}^r) - \sum_d g_p^d(p, Q, a^d)$$
$$- \sum_r g_p^r(p, Q, a^r + F^r) = 0, \quad (5)$$

$$\sum_d F^d - \sum_r F^r = 0. \quad (6)$$

Equations (3) and (4) are the budget constraints for the representative consumer in either a donor or recipient country. They reflect the possibility of exogenously determined transfers to fund

[2]This follows from differentiating the duality relationship $e_p^n(p, U^n) = c_v^n(p, m^n) = c_v^n(p, e^n(p, U^n))$ with respect to U^n, where m^n is the income of country n.

local adaptation measures on the one hand and expenditure for own adaptation on the other. Equation (5) is the market equilibrium condition for the vulnerable good. Because of Walras' law, the market equilibrium condition for the non-vulnerable good can be omitted. Condition (6) finally says that the donors' contributions must cover the total of all transfers.

Now, let us turn to the question, which is in the center of the present analysis: Seen from a purely economic perspective, under which conditions that a donor country has an incentive to contribute to adaptation funding? And conversely, has a recipient country in any case an incentive to accept adaptation funding? For answering this question, let us isolate the effects of transfers for funding adaptation, i.e., suppose that the global climate does not change ($dQ = 0$) and that any change in adaptation measures completely depends on financial transfer ($da^n = 0$). For example, this could be the case if the donors are optimally adapted to climate change, but adaption expenditure in the recipient countries solely depends on transfers.

Through taking the total differentials of conditions (3) and (4), respectively, it becomes obvious that a donor country d has an incentive to increase his contribution to adaptation funding, if

$$dU^d = (g_p^d - e_p^d)dp - dF^d \geq 0, \tag{7}$$

whereas a recipient country r will accept additional funding of adaptation only, if

$$dU^r = (g_p^r - e_p^r)dp + g_a^r dF^r \geq 0. \tag{8}$$

As conditions (7) and (8) show, both the relative trade position, i.e., whether a country is net importer of outputs from vulnerable sectors or not, as well as terms-of-trade effects, i.e., changes of world market prices of vulnerable commodities, matter. Furthermore, the funding decision of a single country strategically affects the decisions of the other countries via prices only.[3] Therefore, let us consider

[3]This is important to note if the public good feature of adaptation funding is considered. If a donor makes sufficient contributions to the global adaptation

how funding adaptation affects the world market prices of vulnerable commodities. From totally differentiating equation (5) as well as conditions (7) and (8) follows:

$$dp = \sum_r \left(\frac{g^r_{pa} - e^r_{pu} g^r_a}{M} \right) dF^r + \sum_d \frac{e^d_{pu}}{M} dF^d, \qquad (9)$$

where $M = \sum_d [e^d_{pp} - g^d_{pp} - e^d_{pU}(e^d_p - g^d_p)] + \sum_r [e^r_{pp} - g^r_{pp} - e^r_{pU}(e^r_p - g^r_p)]$ is the slope of the compensated excess demand function, which is negative if Walrasian stability prevails.

Since $e^n_{pU} > 0$ corresponds to the change in demand of good v in response to a change in income, the second term on the right hand of condition (9) is negative, if the slope of the excess demand function M is negative. Furthermore, $g^r_{pa} - e^r_{pU} g^r_a = (1 - pe^r_{pU}) g^r_{pa}$ measures the compensated marginal effect, which adaptation in a recipient country has on the supply of vulnerable goods. Since pe^r_{pU}, the propensity to consume good v, is smaller than one if v is a normal good, this effect is positive. Hence the first term of condition (9) is also negative, if $M < 0$. In other words, under realistic assumptions it must be expected that increasing the contributions to adaptation funding by at least one donor (and accepting financial support by at least one recipient) implies a reduction of the world market price of vulnerable goods. Condition (9) furthermore indicates that the more it pays to invest into adaptation in a recipient country in terms of net-exports, the higher will be the effect of world market prices of vulnerable commodities.

If a donor country d is net importer of the vulnerable good v, i.e., $g^d_p - e^d_p > 0$, then condition (7) indicates that funding adaptation can, due to terms-of-trade effects, positively affect the donor's welfare. Indeed, being a net-importer of vulnerable commodities is a necessary condition for voluntary funding adaptation. And the higher *ceteris paribus* the dependence on imports of vulnerable commodities, the higher is for strategic reasons the willingness to contribute

---

fund such that the world market prices of vulnerable commodities drop, every net-importer of vulnerable goods can profit.

Table 1. Conditions for Self-Interested Adaptation Funding.

World market prices	$dp = \sum_r \left(\frac{g^r_{pa} - e^r_{pu} g^r_a}{M} \right) dF^r + \sum_d \frac{e^d_{pu}}{M} dF^d < 0$ if $dF^n > 0$ for at least one country n
Donor's contribution	$dF^d > 0$, if $(g^d_p - e^d_p)dp \geq dF^d$. Hence country d must be net importer of vulnerable commodities
Recipient acceptance	$dF^r > 0$, if $g^r_a dF^r \geq (e^r_p - g^r_p)dp$

to adaptation funding. More precisely, gains from price changes set an upper bound on the donor's voluntary additional contributions to adaptation funding, i.e., $(g^d_p - e^d_p)dp \geq dF^d$. That implies in particular that the more a country depends on import of vulnerable commodities and the higher are the terms-of-trade effects, the higher are the voluntary contributions of that country (see Table 1).

A recipient country might also gain from accepting adaptation funding, at least if the reduction of climate change impacts through funded adaptation compensates terms-of-trade deterioration, i.e., $g^r_a dF^r \geq (e^r_p - g^r_p)dp$ (see condition (8)).

3. The Numerical Model

To analyze if and how, depending on trade patterns and projected climate change impacts, adaptation funding in foreign countries is in the self-interests of donor countries, MITACC (Model for Evaluating International Trade and Regional Adaptation to Climate Change) has been applied. MITACC is a dynamic multi-sectoral and multi-regional computable general equilibrium model, based on the model used in Schenker [2013] and further developed in order to explicitly include autonomous investment into adaptation as described in Schenker and Stephan [2014]. MITACC combines a model of world trade with a representation of climate impacts as used in integrated assessment models such as RICE [Nordhaus, 2010] and MERGE [Manne *et al.*, 1995]. In contrast to the former ones, MITACC includes a detailed representation of the regions' ability to adapt to climate change.

In line with the analytical model in Section 2, the world economy now is divided into 13 regions (see Appendix for a list and description of the regional composition). Since climatic impacts depend on the geographic location, while economic wealth influences the ability to adapt against those impacts, countries are assigned to regions according to two criteria: (1) geographic location, and (2) economic wealth, based on data of the World Bank [2013]. This implies that some regions like North America (NOA_high) are characterized by high wealth and thus high adaptive capacities, whereas other ones such as low income Sub-Sahara Africa (SAF_low) or low income South Asia (SOA_low) are less developed and have lower capacities to adapt.

In contrast to the theoretical model, which uses a static framework and neglects intertemporal aspects, the numerical analysis is based on a fully dynamic modeling framework. We suppose that regions follow a Ramsey growth path, i.e., each region acts as if it is represented by a single agent who maximizes the discounted utility of consumption over time. The time horizon of the model is from 2010 to 2075, periods are five years in lengths and the regions' final output can be consumed, invested or put as intermediate input in the production.

Similar to the analytical model, two categories of produced commodities are traded on open international markets: (1) outputs from sectors, which are vulnerable to climate change, such as agriculture, and (2) a composite of outputs from sectors that are almost insensitive to climate change. Production is characterized by nested constant elasticity of substitution (CES) functions with constant returns to scale. Inputs into production are capital, labor and a region-specific intermediate good. We assume factor mobility within but not across regions, i.e., the model does not allow for international capital mobility.[4]

[4]Allowing for capital mobility would open an additional channel through which the world economy as a whole could reduce its exposition by shifting capital to less vulnerable regions. However, individual regions could be hit much harder if

As in the theoretical analysis above, let v denote the vulnerable sector and w the aggregate of sectors, which are not affected by climate change impacts. y_{it}^n symbolizes the gross output of sector $i = v, w$ in region n and period t in absence of climate impacts and adaptation measures. Thereby

$$y_{it}^n = \left(\vartheta_i^n m_{it}^{n\rho_i^y} + (1 - \vartheta_i^n)\Phi_t^n \left[\eta_i^n k_{it}^{n\rho_i^k} + (1 - \eta_i^n) l_{it}^{n\rho_i^k} \right]^{\frac{\rho_i^y}{\rho_i^k}} \right)^{\frac{1}{\rho_i^y}}, \quad (10)$$

with $\rho_i^y = \frac{\sigma_i^y - 1}{\sigma_i^y}$ and $\rho_i^k = \frac{\sigma_i^k - 1}{\sigma_i^k}$. σ_i^y and σ_i^k denote the elasticity of substitution between the respective inputs which are taken as identical across regions. k_{it}^n and l_{it}^n are capital and labor inputs and m_{it}^n is the input of intermediate goods into production of sector i. ϑ_i^n and η_i^n are share parameters and Φ_t^ndenotes the total factor productivity parameter, which is region-specific and exogenous. These parameters differ across regions. Their calibration is explained at detail in Section 3.2.

3.1. *Making adaptation explicit*

As in Section 2, we focus on market damages only and assume that impacts of climate change solely materialize as output losses of vulnerable production. Impacts of climate change can be reduced through adaptation, which, however, creates costs. Most integrated assessment models such as RICE [Nordhaus, 2010] as well as climate impact assessment studies such as PESETA [Ciscar *et al.*, 2011] consider damages, costs and benefits of adaptation only implicitly. However, in order to formulate adaptation as a decision variable both have to be made explicit. Following de Bruin *et al.* [2009], we separate impacts into their constituent elements by assuming: (1) gross

capital would flow to less vulnerable regions. Restrictions on capital mobility in the short run imply that balance of trade must be equalized in every period. However, regional capital stocks can adjust in the long term through changes in net investment (capital stocks can depreciate in one region and accumulate in another).

impacts as well as costs and benefits of adaptation are independent and separable of each other, and (2) all terms can be expressed in units of the affected goods. More precisely, let $\bar{\theta}_t^n$ denote gross impacts of climate change on sector v, i.e., the impacts in absence of adaptation, and let $0 \leq al_t^n \leq 1$ denote the endogenously chosen level of adaptation, which is measured as fraction of gross damages avoided. Then, in region n at time t residual damages, i.e., damages, which remain after adaptation, are

$$\theta_t^n = (1 - al_t^n)\bar{\theta}_t^n. \tag{11}$$

Net output $\tilde{y}_{vt}^n$ of sector v, which is output net of climate impacts and net of adaptation expenditures, then is given by[5]

$$\tilde{y}_{vt}^n = \frac{(1 - a_t^n)}{(1 + \theta_t^n)} y_{vt}^n, \tag{12}$$

where a_t^n denotes, as in the theoretical considerations, costs of adaptation, which are strictly increasing with the level of adaptation

$$a_t^n = \gamma_1 al_t^{n\gamma_2}. \tag{13}$$

γ_1 is the price per unit of adaptation and γ_2 determines the curvature of the adaptation cost function. Both parameters are determined by calibration as is explained in the next section. Note that in the theoretical model, adaptation costs have been expressed in terms of GDP. Here they are expressed in units of the vulnerable good. Similar to the theoretical model we assume that if regions own sufficient resources for optimally investing into adaptation, they choose a level of adaptation such that marginal benefits of adaptation are equal to marginal adaptation expenditures. Thus,

$$a_t^n = \gamma_1 \left(\frac{\gamma_1 \gamma_2}{\bar{\theta}_t^n} \right)^{\frac{\gamma_2}{1-\gamma_2}}. \tag{14}$$

---

[5]Note that production in sector w is by assumption not affected by climate change, hence $\tilde{y}_{wt}^n = y_{wt}^n$.

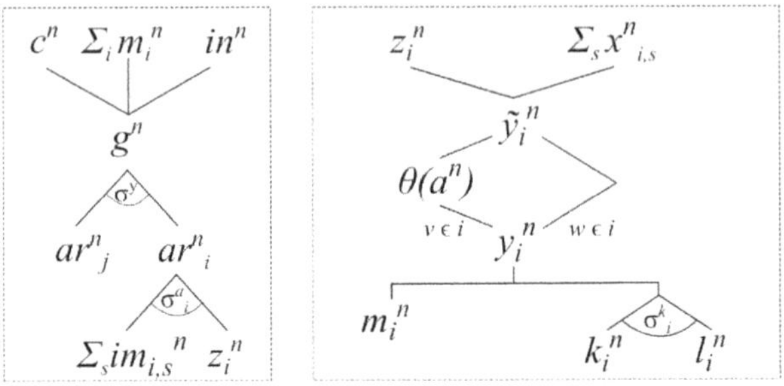

Figure 1. Structure of the Model.

Note: Gross output y_v^n of sector v is exposed to climate impacts $\theta(a^n)$, which depend on adaptation. Net output can be exported $(x_{i,s}^n)$ to region s or used domestically (z_i^n). $im_{i,s}^n$ denotes the import of good i by region n from s. The Armington aggregates ar_i^n are combined to a consumption good g^n, which can be consumed (c^n), used as an intermediate input (m_i^n) in sector i, or invested (in^n) into the regional capital stock.

The rest of the model is similar to the one outlined in Schenker [2013]. As common in most CGE models, we abstract from explicit trade distortions, transportation costs or tariffs and apply Armington's [1968] assumption, which means that imports and domestic products are imperfect substitutes. Figure 1 graphically presents the MITACC's structure. All model equations can be found in the Appendix.

3.2. *Calibration*

3.2.1. *Climate data*

Two variables are important for the assessment of climate impacts on vulnerable sectors such as agriculture. These are changes in temperature and precipitation. We use data provided by the global climate model ECHAM5 [Roeckner *et al.*, 2005] and use simulations which are based on the IPCC SRES A1B emission scenario [Nakicenovic and Swart, 2000]. A1B describes a future world of rapid economic growth and convergence among regions with increased international mobility and a substantial reduction in regional differences in per capita income. Note that the emission scenario does not assume the implementation of any explicit emission reduction measures or policies. The global climate model simulations deliver monthly averages of both temperature and precipitation at a spatial resolution of about 200 km. Through aggregating these data, we get projections of annual mean temperature and precipitation per region. Based on that, we observe i.e., that by the middle of the 21st century the low income East Asia Pacific region (consisting, *inter alia*, of Indonesia and Cambodia) has to cope with an increase of average surface temperature of about 2.7°C. And in low income Sub-Sahara Africa, precipitation will raise by 14% in 2050, whereas the already dry Middle East and North Africa must face a further decline of 5% in precipitation.

3.2.2. *Climate change impacts*

Agriculture is the most important sector among the vulnerable. Therefore, we focus on this sector and assume that agriculture is the sole vulnerable production sector. To construct the sectoral impact function, we rely on a cross-section Ricardian impact function as established by Mendelsohn and Schlessinger [1999] and refined by Cline [2007].

The climate response function is a quadratic combination of temperature and precipitation[6] and determines net revenues per hectare. Taking the total differential gives for each region n and period t

$$\hat{\theta}_t^n = 223.2dT_t^n - 15.74T_t^n dT_t^n + 0.063dP_t^n - 0.00052\, P_t^n dP_t^n. \quad (15)$$

$\hat{\theta}_t^n$ denotes the relative output change, which is determined by the annual average regional surface temperature T_t^n (in degrees Celsius), its marginal changes dT_t^n, annual precipitation P_t^n (in millimeters) and its marginal changes dP_t^n.

Function (15) describes the changes in the opportunity cost per hectare. Consequently, $\hat{\theta}_t^n$ must be interpreted as net-impact, which implicitly includes both the reduction in impacts due to adaptation and their costs. However, for decomposing net-impacts into costs of adaptation a_t^n and residual damages information about gross-impacts and adaptation is needed. Therefore, let us assume that net-impacts

[6] Note this is in contrast to most integrated assessment models, which view net impacts of climate change as function of mean temperature only.

Table 2. Calibrated Parameters of Gross Impact of Climate Change and of Adaptation Cost Parameters.

α_1	α_2	α_3	α_4	γ_1	γ_2
0.340	0.046	0.001	3.806E-6	1.803	3.801

$\hat{\theta}_t^n$ and gross-impacts $\bar{\theta}_t^n$ share the same functional form, i.e.,

$$\bar{\theta}_t^n = \alpha_1 dT_t^n - \alpha_2 T_t^n dT_t^n + \alpha_3 dP_t^n - \alpha_4 P_t^n dP_t^n. \tag{16}$$

The impact coefficients $\alpha_1, \alpha_2, \alpha_3$ and α_4 are then determined together with the adaptation cost parameters γ_1 and γ_2 through calibration. To this end, a model is constructed (for a more detailed description of the calibration procedure, see de Bruin *et al.*, 2009), which minimizes the squared difference between the impacts $\hat{\theta}_t^n$ as given in condition (15), and net-impacts θ_t^n, which derive from the assumption that regions are optimally adapted to climate change impacts. To obtain net-climate impact functions, these authors incorporate adaptation decisions, which can be made on a farm level. This includes instantaneous measures such as shift of crop mixes, changes in fertilizer use, irrigation, and timing of planting and harvesting, where costs and benefits coincide. Table 2 shows the calibrated parameters of the gross-impact function $\bar{\theta}_t^n$ and of the adaptation cost function a_t^n.

3.2.3. *Economic data*

In order to calibrate population growth and technological change, data from the RICE model [Nordhaus, 2010] are used. These projections assume that population growth in the recent future is still positive in most regions except from middle income Europe and Central Asia (ECA_mid) and high income East Asia and Pacific (EAP_high). But from 2060 onwards, the population is declining in almost every region except for Sub-Saharan Africa, North America, and the Middle East.

Regional outputs, production structures, consumption and investment, factor endowments and trade flows are calibrated with GTAP 7.1 data [Narayanan and Walmsley, 2008]. This data base provides a consistent representation of the 2004 world economy and contains information on trade and production for 113 regions and 57 commodities, which are aggregated accordingly. Elasticities of substitution are also taken from the GTAP 7 data base. They are shown in Table A2 in the Appendix.

4. Results of Numerical Simulations

With the CGE model at hand, we are investigating adaptation funding rates that can be motivated by the economic self-interest of the donors purely and do not depend on additional incentives such as fairness considerations or strategic intensions to facilitate negotiations in other issues such as greenhouse gas (GHG) mitigation. Climate impacts in low income developing countries with low or no even negligible adaptive capacities can affect their exports and thus international prices of vulnerable goods such as agricultural commodities. Thus, net-importing high income countries of these goods may have a motivation to fund adaptation in these developing countries in order to protect their terms-of-trade. But the public good nature of adaptation funding might counteract this rationale. For example, if one net-importer finances adaptation measures that are sufficient for moderating the negative impacts of climate change on agriculture, all net-importers benefit from the lower prices on international markets.

Our theoretical considerations in Section 2 demonstrate that at least in principle such conditions can exist, but it remains an empirical question if they actually do. In order to identify such self-sustaining funding equilibria, we simulated a large number of different states of the world, each with different funding rates by the donor nations. In doing this, we assume that developing regions accept the funding offers unconditionally and do not act strategically. As in Schenker and Stephan [2014], it is further assumed that the high income and middle income regions are able to adapt autonomously optimally such that the marginal costs of adaptation are equal to the marginal benefits from it. But low income regions lack this ability and are not able to adapt to climate change. This is a strong assumption, which does not completely fit reality. It provides, however, a first benchmark for judging the effects of climate change impacts. It also rules out situations where the resources from a donor crowd out own adaptation efforts.

4.1. *Identifying potential Nash equilibria of transfer flows*

A first analysis of the model results reveals that the high income region North America is a net-exporter of vulnerable goods in every model period and under every funding scenario. Thus, it is obvious that it is the dominant strategy for NOA_high not to fund other net-exporters.

In the following, we focus only on scenarios where the two remaining developed regions "High income countries in Europe and Central Asia" (ECA_high) and "High income countries in East Asia Pacific region" (EAP_high) each transfer a certain fraction of their Gross National Income (GNI) into an international fund, which then redistributes the resources among low income regions according to the weight of their regional GDP. The recipient regions can use these resources exclusively for adaptation measures.

We ran the CGE model for all possible combinations of funding rates between 0–0.01% of GNI (in steps of 0.001% of GNI) that the two high income donor regions can choose. Similar to Schenker and Stephan [2014], we assume that the funding ramps up and reaches the chosen funding rate from 2035 onwards until the end of the model horizon in 2080. Tables 3a and 3b show the payoffs the two countries get under the different combinations of adaptation funding schemes. The payoffs are expressed in percent of Hicksian Equivalence Variation relative to a scenario, which we call NOFUND. Note that NOFUND represents a world without adaptation funding, where high and middle income regions are optimally adapted to the impacts of climate change, but low income regions are not able to adapt at all. This is a simplified view of the problem but provides

Table 3a. Payoff-Table High Income Europe and Central Asia.

		Funding rates EAP_high										
		0	**0.001%**	**0.002%**	**0.003%**	**0.004%**	**0.005%**	**0.006%**	**0.007%**	**0.008%**	**0.009%**	**0.010%**
Funding rates ECA_high	0	0.0000	*0.0034*	**0.0040**	*0.0042*	*0.0043*	*0.0043*	*0.0043*	*0.0043*	*0.0042*	*0.0041*	*0.0040*
	0.001%	**0.0018**	*0.0017*	*0.0015*	*0.0013*	*0.0011*	*0.0009*	*0.0007*	*0.0005*	*0.0003*	*0.0001*	*−0.0001*
	0.002%	−0.0025	−0.0028	−0.0030	−0.0032	−0.0035	−0.0037	−0.0040	−0.0042	−0.0044	−0.0047	−0.0049
	0.003%	−0.0075	−0.0077	−0.0080	−0.0082	−0.0085	−0.0087	−0.0090	−0.0093	−0.0095	−0.0098	−0.0100
	0.004%	−0.0126	−0.0129	−0.0132	−0.0134	−0.0137	−0.0140	−0.0142	−0.0145	−0.0148	−0.0151	−0.0153
	0.005%	−0.0179	−0.0182	−0.0185	−0.0188	−0.0190	−0.0193	−0.0196	−0.0199	−0.0202	−0.0204	−0.0207
	0.006%	−0.0233	−0.0236	−0.0239	−0.0242	−0.0245	−0.0248	−0.0250	−0.0253	−0.0256	−0.0259	−0.0262
	0.007%	−0.0288	−0.0291	−0.0294	−0.0297	−0.0300	−0.0303	−0.0306	−0.0308	−0.0311	−0.0314	−0.0317
	0.008%	−0.0344	−0.0347	−0.0349	−0.0352	−0.0355	−0.0358	−0.0361	−0.0364	−0.0367	−0.0370	−0.0373
	0.009%	−0.0399	−0.0402	−0.0405	−0.0408	−0.0411	−0.0414	−0.0417	−0.0420	−0.0423	−0.0426	−0.0429
	0.010%	−0.0456	−0.0459	−0.0462	−0.0465	−0.0467	−0.0470	−0.0473	−0.0476	−0.0479	−0.0482	−0.0485

Notes: Payoffs are expressed in percent of Hicksian Equivalence Variation relative to scenario NOFUND. Combinations of simulation runs where adaptation funding leads to welfare gains for the donor are printed in italics. The two resulting Nash equilibra with positive adaptation funding are printed in bold.

Table 3b. Payoff-Table High Income Countries East Asian and Pacific.

		Funding rates EAP_high										
		0	**0.001%**	**0.002%**	**0.003%**	**0.004%**	**0.005%**	**0.006%**	**0.007%**	**0.008%**	**0.009%**	**0.010%**
Funding rates ECA_high	0		*0.0079*	**0.0082**	*0.0079*	*0.0071*	*0.0063*	*0.0052*	*0.0041*	*0.0029*	*0.0017*	*0.0004*
	0.001%	**−0.0006**	*−0.0021*	*−0.0035*	*−0.0050*	*−0.0064*	*−0.0079*	*−0.0094*	*−0.0109*	*−0.0125*	*−0.0140*	*−0.0156*
	0.002%	−0.0173	−0.0189	−0.0204	−0.0220	−0.0236	−0.0252	−0.0268	−0.0285	−0.0301	−0.0317	−0.0333
	0.003%	−0.0353	−0.0369	−0.0386	−0.0402	−0.0419	−0.0435	−0.0452	−0.0469	−0.0485	−0.0502	−0.0519
	0.004%	−0.0540	−0.0556	−0.0573	−0.0590	−0.0607	−0.0624	−0.0640	−0.0657	−0.0674	−0.0691	−0.0708
	0.005%	−0.0730	−0.0747	−0.0764	−0.0781	−0.0798	−0.0815	−0.0832	−0.0849	−0.0866	−0.0883	−0.0900
	0.006%	−0.0923	−0.0940	−0.0957	−0.0974	−0.0991	−0.1009	−0.1026	−0.1043	−0.1060	−0.1077	−0.1095
	0.007%	−0.1117	−0.1135	−0.1152	−0.1169	−0.1186	−0.1204	−0.1221	−0.1238	−0.1256	−0.1273	−0.1290
	0.008%	−0.1313	−0.1331	−0.1348	−0.1366	−0.1383	−0.1400	−0.1418	−0.1435	−0.1452	−0.1470	−0.1487
	0.009%	−0.1511	−0.1528	−0.1545	−0.1563	−0.1580	−0.1598	−0.1615	−0.1633	−0.1650	−0.1667	−0.1685
	0.010%	−0.1709	−0.1726	−0.1744	−0.1761	−0.1778	−0.1796	−0.1813	−0.1831	−0.1848	−0.1866	−0.1883

Notes: Payoffs are expressed in percent of Hicksian Equivalence Variation relative to scenario NOFUND. Combinations of simulation runs where adaptation funding leads to welfare gains for the donor are printed in italics. The two resulting Nash equilibra with positive adaptation funding are printed in bold.

a clear-cut scenario definition that reveals the mechanisms of interest.

The simulation runs identify a number of scenarios where adaptation funding leads to welfare gains for the donors, printed in italics in the payoff tables. However, since in contrast to the analysis in Schenker and Stephan [2014] North America is not forced to participate in such a funding scheme, the resulting funding rates with positive welfare gains are much smaller than in our previous analysis.

We then examine the payoff-table in search for Nash equilibria and scrutinize if a region would have an incentive to deviate from its chosen funding rate given the other regions choice. We can identify two Nash equilibria, i.e., mutual choices of funding rates from which none of the players has an incentive to deviate:

i. High income Europe and Central Asia transfer 0.001% of its GNI into the international fund that supports adaptation measures in low income regions. High income East Asian Pacific region contributes nothing.
ii. High income Europe and Central Asia transfer nothing and high income East Asia Pacific fund adaptation measures in low income countries by 0.002% of its GNI.

It shows that despite its public good character, there are potentially self-sustaining transfer flows from a subset of the developed world to developing countries. These flows are much smaller (more than one order of magnitude) than the amount envisaged under the UNFCCC Long-Term Climate Finance work program.

4.2. *Equilibria under closer inspection*

The first identified Nash equilibrium is nicknamed EQECA after its distinctive characteristic, which is the contribution of the high income Europe and Central Asia to the international adaptation fund. In this equilibrium, ECA_high is willing to transfer 0.001% of its GNI, but remains the only contributor, whereas both other high income regions EAP_high as well as NOA_high are not willing to participate in adaptation funding. The second identified Nash equilibrium is called

EQEAP. It characterizes the situation that ECA_high and NOA_high transfer nothing, but EAP_high funds adaptation in developing countries with 0.002% of its GNI. Both are stable Nash equilibria in the sense that neither ECA_high nor EAP_high have an incentive to deviate from its respective funding rates. Note that in our general equilibrium framework the donating decisions by high income regions affects the welfare of the remaining middle and low income regions. Figure 2a shows welfare effects for the group of high income countries. Figure 2b identifies the consequences for middle income countries, while Figure 2c depicts welfare effects for low income countries.

Figure 2a shows that relative to NOFUND in the EQECA as well as the EQEAP Nash equilibrium, welfare is increasing in both

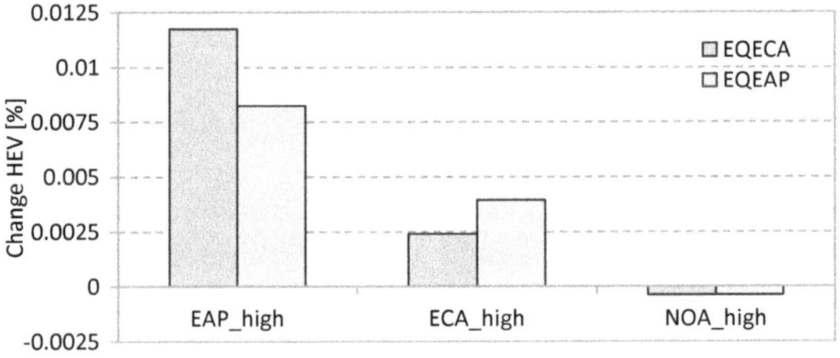

Figure 2a. Percentage Change in Hicksian Equivalence Variation for High Income Regions Relative to a Scenario with No Adaption Funding Scheme.

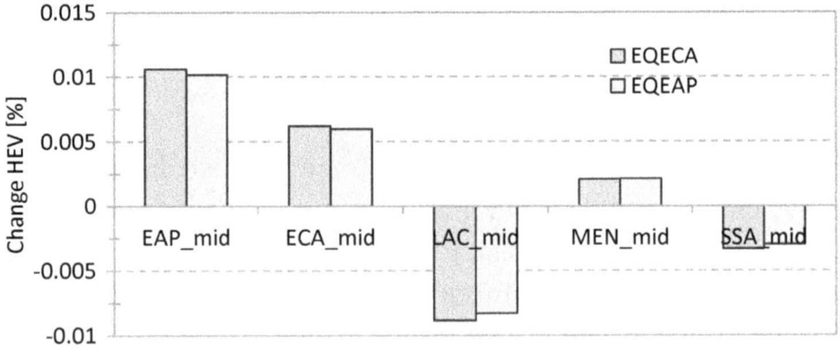

Figure 2b. Percentage Change in Hicksian Equivalence Variation for Middle Income Regions Relative to a Scenario with No Adaption Funding Scheme.

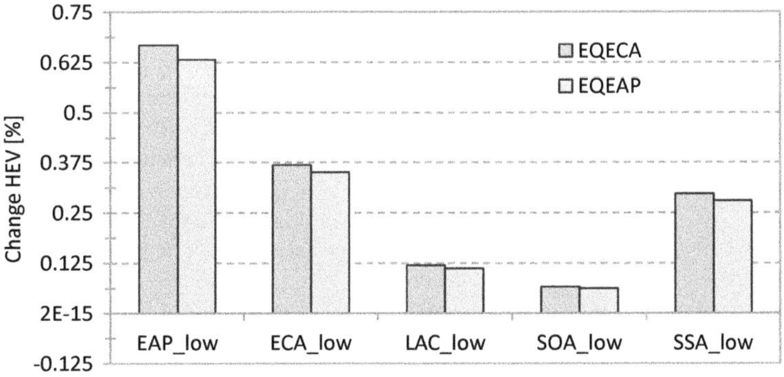

Figure 2c. Percentage Change in Hicksian Equivalence Variation for Low Income Regions Relative to a Scenario with No Adaption Funding Scheme.

high-income net-importers of vulnerable commodities although by a small amount only. Not surprisingly, however the donating region has to bear the main costs, whereas the other one can benefit from its positive effects on its own terms-of-trade. As such, each of these two regions would prefer the other one to take the burden of funding adaptation despite of the fact that if the other one is not acting, acting single handed is still beneficial. Only North America has to cope with small welfare losses if the net-importing high income regions fund adaptation in the vulnerable sector of low income regions. Since the producers of vulnerable goods in the NOA_high region have sufficient resources to autonomously adapt optimally to climate change, a scenario that enables other competitors to improve their productivity in the vulnerable good sector through better adaptation deteriorates NOA_high's comparative advantage in the vulnerable good sector.

Figure 2b depicts the welfare effects for middle income countries. The sign of their welfare effect again depends on the direction of trade in vulnerable goods: If regions are net-importer, they can profit from lower world market prices. However, if a region is net-exporter, such as middle income Latin America, its welfare is decreasing from the worsening of its terms-of-trade.

The largest welfare changes relative to the NOFUND scenario can be observed in the recipient regions. In any case, the pure transfer effect dominates welfare deterioration from terms-of-trade

Table 4. Costs and Benefits of Autonomous Adaptation by 2040 in the Scenario EQECA (Adaptation Levels Are Expressed as Fraction of Damages Avoided in the Vulnerable Sector).

Region	Adaptation costs (billion US $)	Adaptation level (% of gross impacts)	% change in terms-of-trade (rel. to NOFUND)	% change in RCA for vul (rel. to NOFUND)
EAP_high	28.14	40.9%	0.005	−0.27
EAP_low*	0.02	8.6%	−0.244	6.41
EAP_mid	26.64	40.8%	−0.001	−0.26
ECA_high†	1.21	11.8%	0.002	−0.21
ECA_low*	0.04	11.8%	−0.103	10.00
ECA_mid	0.16	12.0%	0.004	−0.23
LAC_low*	0.01	8.1%	−0.004	1.44
LAC_mid	2.06	19.0%	−0.004	−0.15
MEN_mid	0.65	17.3%	0.002	−0.22
NOA_high	1.50	14.4%	−0.006	−0.19
SOA_low*	0.07	8.5%	0.018	1.16
SSA_low*	0.03	8.6%	−0.050	2.63
SSA_mid	0.27	19.0%	−0.005	−0.19

Note: †marks donor regions, *indicates recipient regions.

losses. Note that the magnitude of the welfare gains in the developing regions due to the transfers and the subsequently improved productivity in the vulnerable good sector is several times greater than the welfare effects due to terms-of-trade changes in middle and low income regions.

With respect to its economic effects the two Nash equilibria are very similar to each other. Tables 4 and 5 present key characteristics of the EQECA and EQEAP equilibrium, respectively. The first column in Tables 4 and 5 shows adaptation expenditures in 2040. Total transfers to fund adaptation in low income regions sum up to 160 and 130 million of 2005-USD. This would be much less compared to the envisaged targets in the UNFCCC Long-Term Climate Finance work program. Although some international adaptation funding in developing countries emerges from the self-interest of the high income countries, additional rationales are necessary to increase the volume of contributions as envisaged. Note that relative transfers lead to

Table 5. Costs and Benefits of Autonomous Adaptation by 2040 in the Scenario EQEAP (Adaptation Levels Are Expressed as Fraction of Damages Avoided in the Vulnerable Sector).

Region	**Adaptation costs (billion US $)**	**Adaptation level (% of gross impacts)**	**% change in terms-of-trade (rel. to NOFUND)**	**% change in RCA for vul (rel. to NOFUND)**
EAP_high[†]	28.15	40.9%	0.004	−0.25
EAP_low*	0.02	8.1%	−0.237	5.95
EAP_mid	26.65	40.8%	−0.001	−0.24
ECA_high	1.21	11.8%	0.003	−0.20
ECA_low*	0.03	11.0%	−0.101	9.28
ECA_mid	0.16	12.0%	0.004	−0.21
LAC_low*	0.00	7.6%	−0.004	1.34
LAC_mid	2.06	19.0%	−0.004	−0.14
MEN_mid	0.65	17.3%	0.002	−0.20
NOA_high	1.50	14.4%	−0.006	−0.18
SOA_low*	0.06	8.0%	0.017	1.08
SSA_low*	0.02	8.0%	−0.049	2.44
SSA_mid	0.27	19.0%	−0.004	−0.18

Note: [†]marks donor regions, [*]indicates recipient regions.

only relatively low adaptation levels in the vulnerable good sector of developing regions. The transfer, depicted in the first column of Tables 4 and 5, enable the agricultural sector in Sub-Sahara Africa to avoid 8% of the direct impacts of climate changes (see second column of Tables 4 and 5, respectively). This is below the level in high and middle income regions, where cost-efficient adaptation decisions reduce impacts by 14–40%.

The last column of the Tables 4 and 5, respectively, shows computed changes in the relative competitiveness of the single regions, expressed as the percentage change of the Revealed Comparative Advantage (RCA) Index. The RCA is defined as the share of a region's exports of a commodity in the region's total exports relative to the share of the world's exports of that commodity in global exports (see Balassa, 1965). The RCA shows that adaptation funding can improve the competiveness of the agricultural sector in low-income regions substantially compared to the NOFUND scenario. Low income Sub-Sahara Africa, for example, can improve its competitiveness by about 2% and in low income Europe and Central Asia by even 9–10%.

5. Conclusions

Transfers from the developed to the developing countries will probably be an important cornerstone of every stringent global climate policy agreement. A substantial part of these transfers will be funds for improving the adaptive capacity of developing countries. Most rationales of such North to South transfers are either based on fairness considerations or because transfers are viewed as a strategic device for achieving a stringent and comprehensive agreement on GHG abatement and burden-sharing. Previous research, however, has shown that terms-of-trade effects might be a further justification for adaption funding: Wealthy countries, which are net-importers of climate-sensitive commodities such as agricultural goods, fear the increased scarcity and increasing world market prices due to climate impacts in exporting developing countries. Funding adaptation measures in developing countries can be a cost-efficient insurance against such adverse effects. This we have shown at hand of both a small analytical and large scale numerical model of international trade, which indicates that under certain conditions, countries indeed have an incentive to fund adaptation measures abroad.

There is, however, a further aspect that has to be taken into account. Funding adaptation can have the features of a public good. If adaptation improves the productivity in the vulnerable good sector, this can reduce world market prices of vulnerable goods from which all consumers can profit independently whether they have contributed to adaptation funding or not. Consequently, if there is more than one net-importing country, which has the financial resources to fund adaptation, there is an incentive to free-ride. Using a calibrated multi-region dynamic computable general equilibrium model, we identify Nash equilibria where adaptation funding is part of the strategy. However, the magnitude of North to South transfers in these equilibria is much smaller than what is planned as voluntary contribution to the GCF, which has a target of USD 100 billion. Hence, to reach this goal additional incentives for funding adaptation such as equity and fairness considerations or bargaining rationales that use climate finance to stipulate GHG reduction commitments by recipient countries are necessary. Self-interested actions that are grounded in international trade are not sufficient to close this gap.

We strongly believe that more research on this topic is necessary. For example, an important empirical question is the size of the single contributions. Within our numerical model, regions are relative large such that the action of one player can sufficiently take influence on the terms-of-trade. A further issue in this context is to get better understanding under which conditions different countries would form coalitions for adaptation funding. In the case of high income Europe and Central Asia, which mainly consists of EU countries, such a joint action is not an unrealistic option. But in case of the high income East Asia Pacific regions, which cover countries such as Japan and Australia, differences are too big to believe that coordinated adaptation funding is a realistic policy option. It is thus necessary to understand if and how countries would form adaptation funding coalitions.

Finally, a better understanding of the role governments actually have and what the private sector can contribute, seem necessary. So far, we base our considerations on the assumption that governments are the driving forces. But would not it be reasonable to also assume that firms could also have an interest to fund adaptation in developing countries, in particular if they depend on stable and resilient supply chains to provide the necessary intermediate goods? This could lead to crucial private sector funding since the cash-strapped public sector probably is not able to provide all resources necessary to adapt sufficiently against the threat of climate change.

References

Armington P. 1968. *A Theory of Demand for Products Distinguished by Place of Production.* IMF Working Paper.

Balassa B. 1965. Trade Liberalization and Revealed Comparative Advantage. *Manchester School of Economics and Social Studies* 33, 99–123.

Beutel J. 2002. *The Economic Impact of Objective 1 Interventions for the Period 2000–2006*. Final Report to the Directorate-General for Regional Policies European Commission. http://ec.europa.eu/regional_policy/sources/doc gener/studies/pdf/objective1/final_report.pdf.

Buob S. and G. Stephan. 2013. On the Inventive Compatibility of Funding Adaptation. *Climate Change Economics* 4, 1–18.

Carraro C., J. Eyckmans and M. Finus. 2005. *Exploring the Full Potential of Transfers for the Success of Inter-national Environmental Agreements.* Note di Lavoro 50. Fondazione ENI Enrico Mattei (FEEM). Milan, Italy.

Ciscar J. -C., A. Iglesias, L. Feyen, L. Szabó, D. Van Regemorter, B. Amelung, R. Nicholls, P. Watkiss, O. B. Christensen, R. Dankers, L. Garrote, C. M. Goodess, A. Hunt, A. Moreno, J. Richards and A. Soria. 2011. Physical and Economic Consequences of Climate Change in Europe. *Proceedings of the National Academy of Sciences.*

Cline W. R. 2007. *Global Warming and Agriculture Center for Global Development.* Washington DC: Peterson Institute for International Economics.

CPB Netherlands Bureau for Economic Policy Analysis. 2015. *World Trade Monitor June 2015.* https://www.cpb.nl/en/number/cpb-world-trade-monitor-june-2015.

de Bruin K. C., R. B. Dellink and S. J. R. Tol. 2009. AD-DICE: An Implementation of Adaptation in the DICE Model. *Climatic Change* 95, 63–81.

European Commission. 2013. *An EU Strategy on Adaptation to Climate Change — Impact Assessment Part I.* http://ec.europa.eu/clima/policies/adaptation/what/docs/sw_2013_132_en.pdf

Eyckmans J., S. Fankhauser and S. Kverndokk. 2015. Development Aid and Climate Finance. *Environmental and Resource Economics*, to appear.

Grasso M. 2010. An Ethical Approach to Climate Change Finance. *Global Environmental Change* 20, 74–81.

Hoel M. 2001. *International Trade and the Environment: How to Handle Carbon Leakage.* In Frontiers of Environmental Economics, Folmer *et al.* (eds.), London, UK: Edward Elgar.

Manne A. S., R. Mendelsohn and R. Richels. 1995. MERGE: A Model for Evaluating Regional and Global Effects of GHG Reduction Policies. *Energy Policy* 23, 17–34.

Mendelsohn R. and M. E. Schlesinger. 1999. Climate-Response Functions. *Ambio* 28, 362–366.

Nakicenovic N. and R. Swart (eds.). 2000. *IPCC Special Report on Emissions Scenarios.* Cambridge University Press, Cambridge, UK.

Narayanan G. B. and T. L. Walmsley. 2008. *Global Trade, Assistance, and Production: The GTAP 7 Data Base 2008.* Center for Global Trade Analysis, Purdue University, West Lafayette, IN.

Nordhaus, W. D. 2010. Economic Aspects of Global Warming in a Post-Copenhagen Environment. *Proceedings of the National Academy of Sciences of the United States of America* 107, 11721–11726.

Pittel K. and D. Rübbelke. 2013. International Climate Finance and Its Influence of Fairness and Policy. *World Economics* 36, 419–436.

Roeckner E., G. Bauml and L. Bonaventura. 2005. *IPCC DDC AR4 ECHAM5/MPI-OM SRESA1B run1. World Data Center for Climate. CERA-DB.* Discussion Paper. Max-Planck-Institut für Meteorologie.

Schenker, O. 2013. Exchanging Goods and Damages: The Role of Trade on the Distribution of Climate Change Costs. *Environmental and Resource Economics* 54 (2), 261–282.

Schenker O. and G. Stephan. 2014. Give and Take: How the Funding of Adaptation to Climate Change Can Improve the Donor's Terms-of-Trade. *Ecological Economics* 106, 44–55.

World Bank. 2013. *World Bank National Accounts Data and OECD National Accounts.* Files: GNI Per Capita, Atlas Method (Current US$). Washington, D.C.

Appendix

A.1. Calibration

Table A1. Regional Disaggregation, Calibrated Impact and Adaptation Cost Parameter.

Region	Economies covered	dT_{2050}	dP_{2050}
East Asia Pacific — HI (EAP_high)	Japan, South Korea, Taiwan	2.25	97.29
East Asia Pacific — MI (EAP_mid)	China, Malaysia, Thailand	2.73	34.86
East Asia Pacific — LI	Indonesia, Cambodia, Myanmar, ...	2.69	44.96
Europe and Central Asia — HI (ECA_high)	Germany, France, Italy, Poland, ...	2.36	−4.89
Europe and Central Asia — MI (ECA_mid)	Russian Federation, Bulgaria, ...	3.77	66.06
Latin America — MI (LAC_mid)	Argentina, Brazil, Colombia, ...	2.38	2.07
Latin America — LI (LAC_low)	Paraguay, Bolivia, ...	2.76	−20.25
Middle East and North Africa — MI (MEN_mid)	Egypt, Tunisia, Morocco, ...	2.67	−9.35
North America — HI (NOA_high)	USA, Canada	3.24	55.25
South Asia — LI (SOA_low)	India, Pakistan, Bangladesh, Sri Lanka	2.15	69.01
Sub-Sahara Africa — MI (SSA_mid)	South Africa, Botswana, Mauritius	3.10	−115.1
Sub-Sahara Africa — LI (SSA_low)	Mozambique, Tanzania, Uganda, ...	2.15	14.37

Note: HI stands for high income, MI for middle income, and LI for low income according to World Bank [2013]. Temperature changes are relative to the average of the control period 1961–1990, expressed in degree Celsius. Changes in precipitation relative to the average of the control period 1961–1990 are expressed in millimeter per year.

Table A2. Elasticities of Substitution.

Elasticities of substitution		v	w
Capital — Labor	σ_i^k	0.64	1.31
KL — Intermediates	σ_i^Y	Leontief	Leontief
Armington	σ_i^A	2.42	2.51

A.2. Model Equations of MITACC

Variables and prices

Note that Π^r_{it} denotes the profit function of sector i in region r at time t. Differentiating the profit function with respect to input and output prices provides compensated demand and supply (Shephard's lemma), which appears subsequently in the market clearance condition.

Table A3. List of Activity Levels.

g^r_t	Consumption Good
$\tilde{y}^r_{i,t}$	Gross Output Commodities Production
$y^r_{i,t}$	Net Output Commodities Production
$ar^r_{i,t}$	Armington Good
in^r_t	Investment in r
k^r_t	Capital Stock of Region r
kt^r	Terminal Capital Stock
a^r_t	Adaptation Expenditures as Fraction of Sectorial Gross Output
$F^r_{s,t}$	Additional Adaptation due to Funding in Region r as Fraction of Sectorial Gross Output

Table A4. List of Price Levels.

p^r_t	Price Consumption Good
$p^r_{y,i,t}$	Price Gross Produced Commodities
$p^r_{\tilde{y},i,t}$	Price Net Produced Commodities
$p^r_{a,i,t}$	Price Armington
$p^r_{k,t}$	Price of Assets
$r^r_{k,t}$	Rental Price Capital
w^r_t	Wage Rate

A.2.1. *Zero-profit conditions*

Zero-profit condition in consumption good production:

$$\Pi^r_{Y,t} = p^r_t - \left(\sum_i \theta^r_i {p^r_{a,i,t}}^{1-\sigma}\right)^{\frac{1}{1-\sigma}} = 0 \qquad \text{(A.1)}$$

Zero-profit conditions in commodity production i, gross of climate impact:

$$\Pi^r_{y,i} = p^r_{y,i,t} - \left(\vartheta^r_i p^r_{i,t}{}^{\sigma^y_i} + (1-\vartheta^r_i)\Phi^{r-1}_t[\eta^r_i r^r_{k,t}{}^{\sigma^k_i} + (1-\eta^r_i)w^{\sigma^k_i}]^{\frac{\sigma^y_i}{\sigma^k_i}}\right)^{\frac{1}{\sigma^y_i}}$$

$$= 0, \tag{A.2}$$

where ϕ^r_t denotes exogenous given total factor productivity in region r at time t.

Zero-profit net output (Net of damages, net adaptation expenditures):

$$\Pi^r_{\tilde{y},v,t} = p^r_{y,v,t}\left(\frac{1+\theta^r_t}{1-a^r_t}\right) - p^r_{\tilde{y},v,t} = 0, \quad \Pi^r_{\tilde{y},w,t} = p^r_{y,w,t} - p^r_{\tilde{y},w,t} = 0. \tag{A.3}$$

Zero-profit condition in the Armington production of commodity i:

$$\Pi^r_{a,i,t} = p^r_{a,i,t} - \left(\eta^r_a\, p^r_{\tilde{y},i,t}{}^{1-\sigma^a} \sum_s \eta^r_{A,i,s,t}\, p^s_{i,t}{}^{1-\sigma^a}\right)^{\frac{1}{1-\sigma^a}} = 0. \tag{A.4}$$

Zero-profit condition for capital accumulation:

$$\Pi^r_{k,t} = p^r_{k,t} - (1-\delta)p^r_{k,t+1} - r^r_{k,t} = 0. \tag{A.5}$$

A.2.2. *Market clearance conditions*

Market clearance condition for commodities gross of adaptation costs and climate impacts:

$$\tilde{y}^r_{i,t} = y^r_{i,t}\frac{(1-a^r_t)}{(1+\theta^r_t)}\ \forall v \in i, \quad \tilde{y}^r_{i,t} = y^r_{i,t}\ \forall\, w \in i. \tag{A.6}$$

Market clearance condition for the output of commodity i net of climate impacts:

$$\frac{\partial \Pi^r_{a,i,t}}{\partial p^r_{\tilde{y},i,t}} + \sum_s \frac{\partial \Pi^s_{a,i,t}}{\partial p^r_{\tilde{y},i,t}} = \tilde{y}^r_{i,t}, \tag{A.7}$$

with $\frac{\partial \Pi^r_{a,i,t}}{\partial p^r_{\tilde{y},i,t}}$ and $\frac{\partial \Pi^s_{a,i,t}}{\partial p^r_{\tilde{y},i,t}}$ as domestic demand for commodity i and foreign demand in region s for commodity I, respectively.

Market clearance condition for the consumption good:

$$g_t^r = c_t^r + I_t^r + \sum_i \frac{\partial \Pi_{y,t}^r}{\partial p_t^r} + f_{s,t}^r, \tag{A.8}$$

where $\sum_i \frac{\partial \Pi_{y,t}^r}{\partial p_t^r}$ denotes demand for consumption goods as intermediate input in commodity i production. $f_{s,t}^r$ describes the adaptation funding that flows from r to s.
Market clearance condition for the Armington good i:

$$\frac{\partial \Pi_t^r}{\partial p_{a,i,t}^r} = ar_{i,t}^r, \tag{A.9}$$

with $\frac{\partial \Pi_t^r}{\partial p_{a,i,t}^r}$ as demand for Armington good i.
Market clearance condition for the labor supply:

$$\bar{l}_t^r = \sum_i \frac{\partial \Pi_{y,i,t}^r}{\partial w_t^r}, \tag{A.10}$$

with $\frac{\partial \Pi_{y,i,t}^r}{\partial w_t^r}$ as the demand for labor in sector i.
Market clearance condition for capital demand:

$$k_t^r = \sum_i \frac{\partial \Pi_{y,i,t}^r}{\partial r_{k,t}^r}, \tag{A.11}$$

with $\frac{\partial \Pi_{y,i,t}^r}{\partial r_{k,t}^r}$ as the demand for capital in sector i.
Market clearance condition for investments and capital:

$$k_t^r = k_{t-1}^r(1-\delta) + in_{t-1}^r. \tag{A.12}$$

Household income:

$$h^r = \sum_t w_t^r \bar{l}_t^r + p_{k,0}^r k_0^r - p_{k,T}^r K_T^r. \tag{A.13}$$

Terminal capital condition:

$$\left(\frac{in_T^r}{in_{T-1}^r}\right) = \left(\frac{g_T^r}{g_{T-1}^r}\right). \tag{A.14}$$

A.2.3. Adaptation policy control

Optimal adaptation policy:

$$a_t^r = \gamma_1 \left(\frac{\gamma_1 \gamma_2}{\bar{\theta}_t^r} \right)^{\frac{\gamma_2}{1-\gamma_2}}. \tag{A.15}$$

Residual damages:

$$\theta_t^r = \bar{\theta}_t^r \left(1 - \left(\frac{a_t^r + \sum_d f_{s,t}^d}{\gamma_1} \right)^{\frac{1}{\gamma_2}} \right), \tag{A.16}$$

where $\sum_d f_{s,t}^d$ describes the sum of adaptation support that country r gets from the set of donor countries d.

Chapter 9

Pareto Improvements Induced by Climate Funding in a Strategic Adaptation-Mitigation Framework

Wolfgang Peters

European University Viadrina, Große Scharrnstrasse 59, 15230 Frankfurt (Oder), Germany
peters@europa-uni.de

Reimund Schwarze

Helmholz Center for Environmental Research, Permoserstrasse 15, 04318 Leipzig, Germany
reimund.schwarze@ufz.de

Anna-Katharina Topp

European University Viadrina, Große Scharrnstrasse 59, 15230 Frankfurt (Oder), Germany
topp@europa-uni.de

Abstract

We address the international regime of climate finance, which is emerging in the post-Kyoto architecture, and investigate which type of earmarked funding may lead to Pareto improvements for donor and recipient countries. As funding within the post-Kyoto framework is voluntary, sustained finance in the long term can only be guaranteed if all participating countries benefit. In order to rule out for the Bergstrom paradox, which states that recipient countries may end up in a worse-off situation as a consequence of conditional transfers, we presume a framework in which donor countries commit themselves not to reduce their own mitigation efforts. Regarding three types of earmarked climate funding, which compensate either mitigation, adaptation or damage costs, we find that only funds that are directed at mitigation activities boost the global level of mitigation and may induce Pareto improvements. Transferring our results into the political context of the Green Climate Fund, we recommend to prioritize finance of "energy generation and access", which aims at enhancing mitigation through low-emission power generation and access.

1. Motivation

Over the past years, global negotiations on climate change under the roof of the United Nations Framework Convention on Climate Change (UNFCCC) have stagnated. Only a few countries are willing to commit to meaningful binding greenhouse-gas mitigation target: While 147 parties of the UNFCCC have reported Intended Nationally Determined Contributions (INDCs) to the Paris Agreement (COP21) as of early November 2015, most INDCs are by far not ambitious enough to be capable of meeting the 2°C targets of climate change. However, progress has evolved with respect to the system of climate funding. This new post-Kyoto architecture differs in many respects from the previous Kyoto regime. It links efforts in mitigation to efforts in adaptation and connects them to a comprehensive, worldwide system of monitoring, reporting and verification of greenhouse gas emissions. Moreover, it introduces a new international funding regime of voluntary financial pledges for these efforts, which is the focus of our chapter.

The new architecture for international public climate finance emerging from the COP17 negotiations at Durban, South Africa was preceded by a rudimentary finance architecture of the otherwise largely failed round of negotiations in Copenhagen in 2009 (COP15). The Copenhagen Accord already provides for a fast-track finance (FTF), and mentions (without detailing it) a track of long-term finance (LTF), the then called Copenhagen Green Fund, which was subsequently re-labelled in Cancun as Green Climate Fund (GCF). FTF comprises immediate funding of USD 30 billion in the period

2010–2012 for adaptation and mitigation, but explicitly excludes funding for poverty alleviation and other development objectives, in order to be new and additional to pre-existing funding from the Global Environmental Facility (GEF) of the World Bank and other Official Development Aid (ODA). The contributions to the FTF are voluntary pledges to be communicated to the UNFCCC's secretariat in Bonn. LTF is only mentioned in the Copenhagen Accord as a program to mobilize USD 100 billion per year by 2020 for the immense need for funding adaptation and mitigation measures in developing countries [Barrett, 2008; IPCC, 2007]. The Copenhagen Green Fund was seen as the heart of LTF in the Copenhagen Accord, but the associated call for public and private sources, bilateral and multilateral channels, as well as alternative sources of finance (e.g., insurance and derivatives for climate change loss and damages) already indicates the breadth of LTF in the negotiation context. In the following, we will be concerned with LTF in this comprehensive definition, without mere focus on the Green Climate Fund.

Glemarec [2011] has delivered a useful graphical representation, which depicts the possible new architecture for international public climate finance emerging from the COP17 negotiations (Figure 1). Climate finance will be implemented under the political roof of the UNFCCC's negotiation process (COPs). As such, it will be part of the overall negotiation process. In practice, it will define the total amount of funding, set the rules for contributions and receipt of funding, and select a standing committee on finance to bundle the diverse efforts of LTF. The GCF is at the heart of this funding scheme, but politically connected in transparent ways to other existing UNFCCC funding streams, such as the so-called Adaptation Fund (AF), and parallel to non-UNFCCC streams of funding, such as the GEF of the World Bank. These diverse lines of climate finance will be considered in the COP negotiations holistically, despite the fact that different UNFCCC and non-UNFCCC agencies are in charge of the programming, planning and budgeting of the funding schemes, e.g., the Word Bank's GEF agencies. Mitigation and adaptation shall be considered in the GCF in a balanced manner. A new program, decided in Durban, addresses loss and damage.

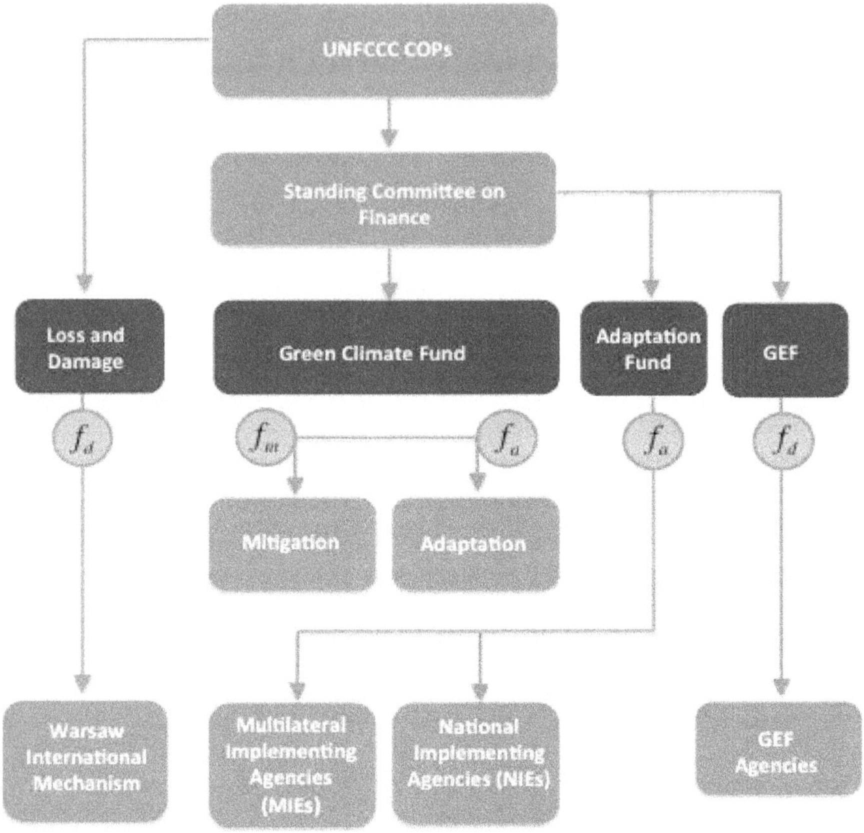

Figure 1. Architecture for International Public Climate Finance after the COP17 Negotiations in Durban [cf. Glemarec, 2011, adjusted].

The present paper analyzes the effects of three different types of climate funding on total costs of climate change in a non-cooperative mitigation-adaptation framework. We regard transfer payments between industrialized countries, who are the donors of the payments, and emerging economies,[1] who are the recipients. In specific, we examine the following three funding instruments, each tied

[1]Although many recipient countries are developing or even least developed countries, here we explicitly regard emerging countries because the latter have a large emission mitigation potential, and thus might contribute to climate protection.

to a different purpose: Funding compensates either a share of the adaptation costs, of the mitigation costs, or of the residual damage costs of the emerging countries. These types of funding instruments reflect the purposes of existing funds presented above; however, we strongly simplify the sophisticated design. Considering these instruments, we investigate how to design the system of climate finance in order to achieve Pareto improvements. Since funding is voluntary, neither donor nor recipient can be forced to participate in funding. Thus, apart from altruistic motives[2] only a funding system which generates Pareto improvements can endure on a long-term basis. As countries negotiate the actual system of climate finance, we model the decision on the levels of the funding instruments as a cooperative decision using the concept of Nash bargaining. The decisions on mitigation and adaptation, on the other hand, are modeled non-cooperatively, since most countries have not agreed so far to binding commitments in climate politics. We find that only funding that compensates a share of mitigation costs in the emerging country yields Pareto improvements since it induces the emerging economy to raise its level of mitigation. As a consequence, the positive externality of mitigation on the industrialized country can be internalized by means of funding. The efficient solution, however, cannot be achieved by unilateral funding as the donor does not internalize the positive externality of its mitigation on the recipient country. Funding that compensates a share of adaptation costs as well as finance that is tied to residual damages do not (necessarily) improve the non-cooperative equilibrium. As a consequence, the result of Nash bargaining rules out funding with regard to adaptation and damages and proposes to only engage in compensating mitigation costs in the emerging country.

Our work is related to the literature on the strategic relationship between adaptation and mitigation as well as to the literature on matching mechanisms, and, more specifically, climate finance. The strategic relationship between adaptation and mitigation has been analyzed within various approaches [see, e.g., Auerswald *et al.*, 2011;

[2]We disregarded altruistic motives in this framework assuming that self-serving motives dominate in governmental decisions.

Ebert and Welsch, 2012; Eisenack and Kähler, 2012; Heuson *et al.*, 2015a; Zehaie, 2009]. While mitigation has the characteristics of a global public good, the benefits of adaptation occur on a private or national level, and is thus modeled as a private good. One central question is whether the strategies of adaptation and mitigation are complements or substitutes with respect to reducing the damage costs of climate change. In case of complements, the desired level of mitigation (adaptation) increases with rising mitigation (adaptation), and in case of substitutes the desired level of mitigation (adaptation) decreases with more adaptation (mitigation). Studies from Ingham *et al.* [2005] and Buob and Stephan [2011] state that in special cases adaptation and mitigation are complements, for example in case of maladaptation, which is a kind of adaptation that causes emissions and, thus, makes further mitigation necessary. Most literature, however, assumes a substitutive relationship since mitigation lowers damage costs, and thus reduces the desired level of adaptation and vice versa [see, e.g., Ebert and Welsch, 2012; Pittel and Rübbelke, 2013]. In this chapter we follow the latter with more general reasoning and focus on mitigation and adaptation as substitutes.

Climate finance can be seen as a unilateral conditional transfer (matching) as it is earmarked for a specific purpose. This type of climate finance alters the recipient's relative costs of mitigation and adaptation and, in turn, influences the non-cooperative decisions. Thus, a well-designed system of climate funding might induce Pareto improvements for donor as well as recipient countries. However, as shown by Bergstrom [1989] and Buchholz *et al.* [2015; 2017], a unilateral matching of public good contributions might lead to the paradoxical effect that the recipient of transfers gets worse off due to offsetting behavior of the donor.[3] We rule out this so-called Bergstrom paradox regarding a framework in which the donor commits to a certain level of mitigation and, accordingly, commits not to offset its contribution to the public good mitigation.

The most closely related works are those from Buob and Stephan [2013], Pittel and Rübbelke [2013] and Heuson *et al.* [2015a] who

[3]For a more detailed description of the effect, see Section 3.1.

investigate the role of adaptation finance. Buob and Stephan [2013] consider a non-cooperative framework in which the industrialized North commits itself to a certain level of mitigation and, in a second stage, decides on domestic adaptation as well as adaptation transfers to the developing South. They find that only if adaptation and mitigation of the South are complements, adaptation finance can be beneficial to the donor country. Moreover, a tighter mitigation commitment of the donor country crowds out adaptation funding. Pittel and Rübbelke [2013] compare the performance of in-kind transfers in form of adaptation that are conditional on the mitigation level of the recipient region with mitigation subsidies. According to Bergstrom [1989], Pittel and Rübbelke [2013] find that both types of conditional transfers may make the recipient worse and the donor region better off. Moreover, conditional adaptation transfers can be more effective than mitigation subsidies, which might explain the existence of adaptation funds. However, this result holds only under specific conditions and the authors conclude that the occurrence of adaptation funds generally remains a puzzle. Heuson *et al.* [2015a] investigate the effects of various modes of the existing system of climate finance setting up minimum requirements for conditional climate transfers to achieve Pareto improvements. They take up a damage function of Kane and Shogren [2000] that distinguishes between the probability of climate change, which can be reduced by mitigation, and the potential damages, which can be lowered by means of adaptation. Heuson *et al.* [2015a] find that subsidizing mitigation as well as transfers which are conditional on the vulnerability of a country fulfills the minimum requirements for a Pareto improvement. The present chapter differs from Heuson *et al.* [2015a] with respect to the following points: First, it takes up a more general damage function, which does not distinguish between the effects of mitigation reducing the probability of climate change and adaptation lowering the potential damage costs. Second, it models the decision on the funding instruments by Nash bargaining, which reflects that funds like the GCF are decided under the political roof of the UNFCCC regime. Third, the present approach identifies actual Pareto improvements instead of minimum requirements. Moreover, this chapter, rules out the Bergstrom paradox establishing a political framework in which the donor country credibly commits to a certain level of mitigation.

The remaining chapter is organized as follows: Section 2 presents the basic adaptation-mitigation model of climate change. Moreover, the non-cooperative benchmark as well as the efficient solution are established. Section 3 introduces the funding instruments and determines the funding equilibrium, which is compared to the benchmark allocations. Section 4 discusses the main results and its possible application.

2. Economic Framework

In this section, we introduce a basic economic model of climate change regarding the decisions on mitigation and adaptation. Moreover, we determine two benchmark allocations, the non-cooperative Nash equilibrium and the efficient allocation, which will serve as points of reference for the assessment of different funding instruments in Section 3.

2.1. *Basic model*

We set up a two-country framework[4] with an industrialized country (uppercase letters) and an emerging country (lowercase letters) following Ebert and Welsch [2012]. Countries decide on their efforts in mitigation of greenhouse gas emissions (M, m) as well as on their levels of adaptation to climate change (A, a). Mitigation and adaptation, in each case, reduce the damage costs of climate change $D(M + m, A)$ with decreasing marginal returns, i.e., $D_1 < 0, D_{11} > 0$ and $D_2 < 0, D_{22} > 0$.[5] While mitigation is a

[4]The two countries may represent in each case a group of industrialized and emerging countries, respectively. We do not regard free-riding incentives within these groups. Instead we focus on the modification of the incentive structure between the two groups due to climate finance.

[5]Moreover, the damage function D is strictly convex, i.e., $D_{11}D_{22} - D_{12} > 0$, which guarantees the fulfillment of each countries' second-order condition as well as well-behaved comparative statics (cf. footnote 10).

public good and, accordingly, reduces damage costs in both countries, adaptation is modeled as a private good, which is only effective on the national level. Furthermore, we assume that adaptation and mitigation are substitutes in the sense that mitigation reduces the effectiveness of adaptation and vice versa, $D_{12} > 0$. In other words, the more mitigation (adaptation) is undertaken, the less adaptation (mitigation) is desired. For example, an increase in mitigation might prevent the sea level from rising and flooding a village, which otherwise could be protected by adaptation in form of a dyke. Costs of efforts in mitigation $C(M)$ and adaptation $G(A)$ are modeled as increasing and convex functions in the respective variables:

$$\begin{aligned} TC &= D(M+m, A) + C(M) + G(A), \\ tc &= d(M+m, a) + c(m) + g(a). \end{aligned} \tag{1}$$

We regard the case in which countries engage simultaneously in mitigation and adaptation. This case is analogous to the sequencing of fixing mitigation before adaptation since in each case strategic incentives do not arise [see, e.g., Heuson *et al.*, 2015b; Zehaie, 2009]. The latter sequencing is usually seen as the natural order since mitigation combats the cause of climate change and adaptation alleviates the consequences of climate change that had not been avoided by mitigation.[6] Thus, we only rule out for the sequencing of advancing adaptation before mitigation.

2.2. *Benchmarks*

In order to assess the effects of funding, we need to establish two benchmark allocations. The "business-as-usual" scenario is represented by the non-cooperative Nash equilibrium of the basic model. In a second step, we determine the efficient solution and compare the non-cooperative equilibrium to the efficiency benchmark in order to determine the scope for Pareto improvements.

---

[6]For a detailed discussion of possible sequencing see, e.g., Ebert and Welsch [2012] who argue that emissions are inseparably tied to economic activity and, therefore, mitigation cannot be chosen after adaptation.

2.2.1. *Non-cooperative equilibrium without funding*

In the Nash-equilibrium without funding, each country minimizes its respective total costs (1) with respect to domestic mitigation and adaptation. The first-order conditions for the emerging country read

$$\frac{\partial tc}{\partial a} = d_2(M + m, a) + g_1(a) = 0, \tag{2}$$

$$\frac{\partial tc}{\partial m} = d_1(M + m, a^N) + c_1(m) = 0. \tag{3}$$

The first-order conditions for the industrialized country are analogous. This system of four equations yields the allocation in the Nash equilibrium without funding (M^N, m^N, A^N, a^N).

2.2.2. *Efficiency benchmark*

By minimizing aggregate total costs with respect to all four decision variables, $\min_{A,a,M,m}\{TC + tc\}$, the efficient allocation arises. The first-order conditions with respect to adaptation are analogous to the non-cooperative ones (2). Regarding mitigation of the emerging country, the first-order condition in case of efficiency reads

$$\frac{\partial TC + tc}{\partial m} = d_1(M + m, a) + c_1(m) + D_1(M + m, A) = 0. \tag{4}$$

The first-order condition with respect to mitigation of the industrialized country is analogous. Comparing (3) with (4), we notice that the positive externality of mitigation in the emerging country on damage costs of the industrialized country, D_1, is internalized in efficiency but not in the Nash equilibrium. The same applies for the industrialized country's level of mitigation. Conversely, within the non-cooperative framework, the problem of underprovision arises. Adaptation, in turn, increases since it serves as a substitute to mitigation.[7]

Accordingly, there is scope for Pareto improvements: The efficient allocation can be approached by an increase in the level of global mitigation (independently of its origin), and a consequential decrease in adaptation in both countries.

[7]For a more detailed analysis of the comparison of the Nash equilibrium with the efficient solution, see Heuson *et al.* [2015b].

3. Funding

In this section, we investigate the effect of climate funding on the global allocation of mitigation and adaptation. In particular, we ask which modes of conditional transfer payments from industrialized towards emerging countries may yield Pareto improvements. First, we discuss the impact of the so-called Bergstrom paradox and frame a strategy to avoid its occurrence. Next, we identify the equilibrium in the last stage, in which countries decide on mitigation and adaptation, and we determine the impacts of funding on the equilibrium. Finally, we model a preceding stage in which the levels of funding instruments are chosen and we assess each instrument separately whether it may induce a Pareto improvement.

3.1. *Bergstrom paradox*

As shown by Bergstrom [1989] and Buchholz *et al.* [2015; 2017], unilateral matching of the public good supply induces a paradoxical effect placing the recipient of the transfer payments into a worse-off situation. The reason is that the donor country incentivizes the recipient to enlarge its contribution to the public good mitigation. However, since the mitigation levels are strategic substitutes, the donor country partially offsets this increase in mitigation by a lower own contribution. As a result, the donor country shifts the burden of mitigation costs towards the recipient country. Purely conditional transfers that are earmarked for additional mitigation (beyond the non-cooperative level) in the emerging economy would evoke this paradox, which hampers a Pareto improvement, in the same way.

In order to rule out this so-called "Bergstrom paradox", we regard a political framework in which the industrialized country credibly commits itself to a certain level of mitigation which is exogenously determined [cf. Buchholz *et al.*, 2015; 2017]. A plausible commitment would be not to deviate from the amount of mitigation in Nash equilibrium $\overline{M} = M^N$, which would also be the "outside option", i.e., the level of mitigation the donor chooses if negotiations on climate funding fail.[8] By fixing mitigation in the industrialized country, damage D does, in turn, only depend on the emerging country's mitigation m, i.e., $D(m, A)$ and $d(m, a)$. This commitment of the donor country could become credible by engaging in up-front investments in mitigative technology. Due to the undertaken expenses, crowding out would not be beneficial to the industrialized country.

3.2. *Set of funding instruments*

Now, we integrate into the basic model a stylized system of climate transfers between the industrialized country, which is the donor of the transfers, and the emerging country, which receives the funding. We distinguish between three different instruments of funding which in each case have a different purpose:

- f^a compensates a share of the adaptation costs of the emerging economy,
- f^d compensates a share of the residual damage costs of the emerging country,
- f^m compensates a share of the mitigation costs of the emerging country.

Accordingly, f^d can be interpreted as vulnerability-related funding, whereas f^a and f^m serve as matching instruments for adaptation and mitigation efforts in the emerging country. We do not include the possibility of negative funding, which would imply taxing the recipient country's mitigation contributions, i.e., $(f^a, f^m, f^d) \in \mathbb{R}^3_+$.[9]

[8] As some countries like the European Union (EU) aim at promising high mitigation contributions, it could also be discussed what happens if the donor country fixes its mitigation effort at a higher level than the amount of the non-cooperative Nash equilibrium. Comparative statics would remain the same as outlined in the following analysis, and the Bergstrom paradox would not apply, either. However, in such a framework, it cannot be guaranteed that the donor country still benefits from funding since it faces two kinds of additional costs: First, it pays the transfers, and second, it also bears the costs of increasing its own mitigation efforts.

[9] These funding instruments do not correspond exactly to the existing tiers of funding under the roof of the UNFCCC since the guidelines of funding are far more sophisticated. Nevertheless, these stylized instruments do catch the basic types of earmarking which persist in UNFCCC funding as outlined in Section 1.

The total costs of climate change of the industrialized and the emerging country, respectively, including climate funding are given by

$$TC^f = D(m, A) + C(\overline{M}) + G(A) + f^m c(m) + f^a g(a) + f^d d(m, a), \tag{5}$$

$$tc^f = \left[1 - f^d\right] d(m, a) + [1 - f^m] c(m) + [1 - f^a] g(a).$$

3.3. *Impact of funding on mitigation and adaptation*

In this subsection, we determine the equilibrium regarding adaptation and mitigation in the last stage of the game and we investigate the impact of the different funding instruments on the equilibrium allocation. In order to keep the analysis compact, we determine the first-order conditions for a Nash equilibrium including all three instruments. However, we analyze each instrument separately and identify whether it might induce a Pareto improvement.

The first-order conditions of the emerging and the industrialized country regarding adaptation read

$$\frac{\partial tc^f}{\partial a} = \left[1 - f^d\right] d_2(m, a) + [1 - f^a] g_1(a) = 0, \tag{6}$$

$$\frac{\partial TC^f}{\partial A} = D_2(m, A) + G_1(A) = 0. \tag{7}$$

Moreover, the emerging country decides on mitigation m while the level of mitigation of the industrialized country is fixed to the level in the non-cooperative Nash equilibrium $\overline{M} = M^N$. The first-order condition with respect to mitigation in the emerging economy reads

$$\frac{\partial tc^f}{\partial m} = [1 - f^d] d_1(m, a) + [1 - f^m] c_1(m) = 0. \tag{8}$$

Equations (6), (7) and (8) determine the equilibrium of the last stage of the sequential game (m^f, A^f, a^f). The decision on adaptation in the industrialized country (7) does not have any implications on

Therefore, this analysis might be interpreted as a basic evaluation of different modes of earmarking funds for emerging countries.

decisions within the emerging economy. Thus, the comparative statics of this equilibrium with respect to the funding instruments can be established by means of (6) and (8) only.

We determine the comparative statics of the equilibrium in the last stage by totally differentiating the system of equations (6) and (8) with respect to the three funding instruments f^a, f^d and f^m. Rearranging terms, comparative statics for the case of funding instrument f^a yield

$$\frac{dm}{df^a} = \frac{-g_1\left[1-f^d\right]d_{12}}{det} < 0, \qquad \frac{da}{df^a} = g_1\frac{\left[1-f^d\right]d_{11}+\left[1-f^m\right]c_{11}}{det} > 0.^{10} \tag{9}$$

Since adaptation becomes cheaper due to the partial compensation of adaptation costs, the level of adaptation of the emerging economy increases in equilibrium due to instrument f^a. As a consequence, the emerging country substitutes mitigation with adaptation. Thus, the effect of instrument f^a on the emerging country's level of mitigation in equilibrium is negative.

With regard to instrument f^d, comparative statics yield

$$\frac{dm}{df^d} = \frac{1}{det}\left[\underbrace{d_1\left[\left[1-f^d\right]d_{22}+\left[1-f^a\right]g_{11}\right]}_{\text{direct eff.}(-)}\underbrace{-d_2\left[1-f^d\right]d_{12}}_{\text{indirect eff.}(+)}\right],$$
$$\frac{da}{df^d} = \frac{1}{det}\left[\underbrace{d_2\left[\left[1-f^d\right]d_{11}+\left[1-f^m\right]c_{11}\right]}_{\text{direct eff.}(-)}\underbrace{-d_1\left[1-f^d\right]d_{12}}_{\text{indirect eff.}(+)}\right]. \tag{10}$$

[10]The strict convexity of the damage function $d(\cdot)$ is a sufficient condition for det being positive: $det = [[1-f^d]d_{22}+[1-f^a]g_{11}][[1-f^d]d_{11}+[1-f^m]c_{11}]-[[1-f^d]d_{12}]^2 > 0$.

In case of instrument f^d, two opposing effects arise. A partial compensation of the damage costs f^d induces a decrease in mitigation as well as in adaptation of the emerging country, which is the direct effect of instrument f^d. The reason is that the emerging country faces lower costs of climate change and, accordingly, mitigation as well as adaptation become less effective. However, due to the substitutional relationship between mitigation and adaptation, the decrease in mitigation induces an indirect positive effect on adaptation and vice versa. If the substitutional effect between mitigation and adaptation is small, i.e., $d_{12} \to 0$, the indirect effects disappear. As a consequence, the direct effects outweigh the indirect ones and a partial compensation of damages lowers the levels of mitigation and adaptation, i.e., $\frac{dm}{df^d} < 0$ and $\frac{da}{df^d} < 0$.

Comparative statics of the equilibrium in the last stage with respect to funding instrument f^m yields

$$\begin{aligned} \frac{dm}{df^m} &= \frac{c_1\left[\left[1-f^d\right]d_{22}+\left[1-f^a\right]g_{11}\right]}{det} > 0, \\ \frac{da}{df^m} &= \frac{-c_1\left[1-f^d\right]d_{12}}{det} < 0. \end{aligned} \tag{11}$$

Instrument f^m, which compensates a fraction of the mitigation costs of the emerging economy, induces an increase in the level of mitigation of the emerging country. This, in turn, negatively affects adaptation in the emerging country. The reason is that due to instrument f^m, mitigation in the emerging country becomes cheaper; as a consequence, the emerging country partially substitutes adaptation for mitigation.

While instruments f^a and f^d fail to raise the global level of mitigation, instrument f^m has a positive effect on mitigation and, thus, may induce a Pareto improvement. The next section illustrates how the levels of funding are decided within Nash bargaining.

3.4. *Nash bargaining with side payments*

We assume that countries decide on the levels of the funding instruments in a first stage before fixing mitigation and adaptation. As the members of the GCF originate equally from emerging and industrialized countries, it seems appropriate to model the decision on the size of the funding instruments by Nash bargaining. For reasons of simplicity, we assume that countries can split the efficiency gains of Nash bargaining by means of a side payments s.[11]

In stage 1, the industrialized and the emerging country negotiate the level of the respective funding instruments f^a, f^m, f^d as well as the transfer payment s. The outside options of the emerging and the industrialized countries are given by its respective total costs that occur in case of non-cooperative Nash behavior (see Section 2.2), i.e., $tc^N = tc(M^N, m^N, a^N)$ and $TC^N = TC(M^N, m^N, A^N)$. The objective function of Nash bargaining reads

$$B = \left[TC^f - s - TC^N\right]\left[tc^f + s - tc^N\right] \rightarrow \max_{f^a, f^d, f^m, s}. \tag{12}$$

Since we allow for side-payments s, the cost savings due to funding are split equally among the two countries. With respect to the funding instruments, the following Kuhn-Tucker conditions arise

$$f^\bullet \geq 0 \text{ and } f^\bullet \left[\frac{\partial TC^f}{\partial f^\bullet} + \frac{\partial tc^f}{\partial f^\bullet}\right] = 0, \tag{13}$$

where $\bullet = a, m, d$, respectively. Thus, the levels of funding are chosen in order to minimize aggregate total costs, i.e., $\min\{TC^f + tc^f\}$, which, in turn, guarantees a second-best efficiency. For institutional reasons, $f^\bullet$ is restricted to be non-negative as we deal with funding emerging countries through financial aids by industrialized countries.

Evaluating each of the three funding instruments separately, i.e., fixing the remaining funding instruments to zero, we receive the following results.

[11] As we allow for side payments, efficiency gains are always split equally among the two bargaining parties. Consequently, the Nash bargaining solution is simplified and it is possible to analyze the bargaining outcome more profoundly. However, this is only an assumption and other splits are possible as well.

i) Funding adaptation f^a

The first-order condition with respect to instrument f^a with $f^d = f^m = 0$ is given by

$$f^a \left[\frac{\partial TC}{\partial f^a} + \frac{\partial tc}{\partial f^a} \right] \Bigg|_{f^m = f^d = 0}$$

$$= f^a \left[\underbrace{\frac{\partial m}{\partial f^a} D_1 + \frac{\partial a}{\partial f^a} g_1 f^a}_{+} \right] = 0. \qquad (14)$$

A positive f^a increases aggregate total costs from climate change. This can be intuitively explained by the fact that f^a, which partially compensates adaptation costs, has a negative effect on the level of mitigation in the emerging country (see (9)). Thus, as the term in square brackets is positive, the optimal f^a equals zero.

ii) Funding according to damage costs f^d

For instrument f^d, the first-order condition reads

$$f^d \left[\frac{\partial TC}{\partial f^d} + \frac{\partial tc}{\partial f^d} \right] \Bigg|_{f^a = f^m = 0}$$

$$= f^d \left[\underbrace{\frac{\partial m}{\partial f^d} \left[D_1 + f^d d_1 \right] + \frac{\partial a}{\partial f^d} d_2 f^d}_{+} \right] = 0. \qquad (15)$$

Again, we observe that f^d raises aggregate total costs of climate change at least for the case that the direct effect outweighs the indirect one (see (10)). The reason is that instrument f^d partially compensates the residual damage costs. This, in turn, lowers the benefits of mitigation. In order to reduce total costs, however, the level of mitigation needs to rise. Thus, countries optimally do not engage in funding according to residual damages, i.e., $f^d = 0$.

iii) Funding mitigation f^m

Considering instrument f^m, the first-order condition reads

$$f^m \left[\frac{\partial TC}{\partial f^m} + \frac{\partial tc}{\partial f^m} \right] \Bigg|_{f^a = f^d = 0} = f^m \frac{\partial m}{\partial f^m} \left[\underbrace{D_1}_{-} + \underbrace{f^m c_1}_{+} \right] = 0. \tag{16}$$

Solving for the optimal level of f^m yields

$$f^m = \frac{-D_1}{c_1} > 0. \tag{17}$$

The optimal f^m is positive since it induces an increase in the level of mitigation in the emerging country. Due to this type of funding, the positive externality of mitigation can be (at least partially) internalized and, thus, it serves as an instrument for Pareto improvements.

Inserting the optimal level of funding $f^m = \frac{-D_1}{c_1}$ into the first-order conditions of the emerging country (8) and (6), and refraining from the instruments $f^a = f^d = 0$ yields

$$d_1 + g_1 = 0, \tag{18}$$

$$D_1 + G_1 = 0, \tag{19}$$

$$d_1 + c_1 + D_1 = 0. \tag{20}$$

These three first-order conditions correspond to the efficient solution outlined in Section 2.2.2. The positive externality of mitigation of the emerging economy on the industrialized country is internalized. However, as the level of mitigation in the industrialized country is fixed at $\overline{M} = M^N < M^*$, the first-best solution cannot be implemented via funding, and, instead, a second-best arises.

4. Conclusion

Climate change negotiations on binding emission targets have not been successful for several years. The only measure countries agreed upon are various climate funds which support developing and emerging countries in mitigation and adaptation activities as well as in compensating damages. These funds might encourage recipient countries to extend their mitigation activities and, thus, alleviate the problem of underprovision of mitigation. Moreover, the existing funding architecture can only be sustained on a long-term basis if donor as well as recipient countries benefit.

We examined three different climate funding instruments and their impacts on the non-cooperative decisions on mitigation and adaptation. In the framework of this simplified model, funding may create Pareto improvements only if transfer payments increase the level of mitigation in the recipient country. We rule out for the occurrence of the Bergstrom paradox which states that transfers may harm the recipient country due to offsetting behavior of the donor country. Considering funding that either compensates adaptation, damage or mitigation costs, we find that only the latter instrument may induce Pareto improvements. The reason is that due to funding, mitigation in the recipient country becomes cheaper and, thus, the emerging economy increases its level of mitigation. Compensating a share of the adaptation costs has the analogous effect on adaptation in the recipient country. However, since adaptation and mitigation are substitutes in reducing the costs of climate change, an increase in adaptation induces a decline in mitigation. Thus, a Pareto improvement is not possible with adaptation funding. Transfer payments which compensate a share of the damage costs in the emerging country reduce the losses of climate change, and, consequently, make adaptation and mitigation less beneficial. If this direct effect predominates, mitigation declines and Pareto improvements due to funding are not achievable.

If funding according to adaptation costs and damage costs, respectively, does not yield Pareto improvements, it is not in line with the interest of both recipient and donor country. Therefore, it is questionable how those funds will be financed voluntarily in the future. However, our approach does not include indirect benefits from funding and other-regarding preferences. Funding in terms of damage costs and adaptation supports a sustainable development of emerging economies. Therefore, donor countries may benefit in the long term from emerging economies that grow and prosper. Moreover, our approach does not take into account motives of fairness, reciprocity or altruism, which also might induce donor countries to engage in international climate funding.

Within our model framework, funding mitigation is the only type of climate finance that can guarantee Pareto improvements of donor and recipient countries. However, although funding of mitigation costs in emerging countries may alleviate the problem of climate change, it is not possible to achieve the first-best allocation via unilateral funding. The reason is also that donor countries need to increase their levels of mitigation apart from rendering the financial means of funding. Thus, unilateral funding might render a second-best solution and alleviate the damages of climate change. However, binding commitments of donor countries are also necessary to combat climate change.

Transferring these results carefully into the political context, it is recommendable to prioritize the funding of mitigation activities. Accordingly, projects associated with the "energy generation and access" within the GCF, which aim at enhancing mitigation regarding low-emission electricity generation and energy use [GCF, 2015: 12], should be promoted. The current system of "investment priorities" according to the GCF-board [GCF, 2015: 9], however, does not lean towards a structure of funding that would allow to establish a system of climate finance that induces Pareto improvements and, thus, could be sustained in the long term.

References

Auerswald, H., K.A. Konrad and M. Thum. 2011. Adaptation, Mitigation and Risk-Taking in Climate Policy. Tech. rep., CESifo working paper: Energy and Climate Economics.

Barrett, S. 2008. Climate Treaties and the Imperative of Enforcement, *Oxford Review of Economic Policy* 24 (2), 239–258.

Bergstrom, T. 1989. Puzzles: Love and Spaghetti, the Opportunity Cost of Virtue. *The Journal of Economic Perspectives*, 165–173.

Buchholz, W., R. Cornes, W. Peters and D. Rübbelke. 2015. Pareto Improvement through Unilateral Matching of Public Good Contributions: The Role of Commitment. *Economic Letters* 132, 9–12.

Buchholz, W., R. Cornes, W. Peters and D. Rübbelke. 2017. Funding International Climate Protection: How to Overcome the Bergstrom Paradox?, in "Climate Finance: Theory and Practice", 213–228.

Buob, S. and G. Stephan. 2011. To Mitigate or to Adapt: How to Confront Global Climate Change. *European Journal of Political Economy* 27 (1), 1–16.

Buob, S. and G. Stephan. 2013. On the Incentive Compatibility of Funding Adaptation. *Climate Change Economics*, 4 (2).

Ebert, U. and H. Welsch. 2012. Adaptation and Mitigation in Global Pollution Problems: Economic Impacts of Productivity, Sensitivity, and Adaptive Capacity. *Environmental and Resource Economics* 52 (1), 49–64.

Eisenack, K. and L. Kähler. 2012. Unilateral Emission Reductions Can Lead to Pareto Improvements When Adaptation to Damages is Possible. *University of Oldenburg, Department of Economics Working Papers* 344 (12).

Green Climate Fund (GCF). 2015. Analysis of the Expected Role and Impact of the Green Climate Fund. Tech. rep. GCF/B.09/06. GCF.

Glemarec, Y. 2011. Catalysing Climate Finance: A Guidebook on Policy and Financing Options to Support Green, Low-Emission and Climate-Resilient Development. Tech. rep. United Nations Development Programme.

Heuson, C., W. Peters, R. Schwarze and A.K. Topp. 2015a. Voluntary International Climate Finance under the Post-Kyoto Framework: The Strategic Consequences of Different Modes of Funding. *Climate Change Economics* 6 (3).

Heuson, C., W. Peters, R. Schwarze and A. K. Topp. 2015b. Investment and Adaptation as Commitment Devices in Climate Politics. *Environmental and Resource Economics.*

Ingham, A., J. Ma and A.M. Ulph. 2005. How Do the Costs of Adaptation Affect Optimal Mitigation When There Is Uncertainty, Irreversibility and Learning. *Tyndall Centre for Climate Change Research Report* 74.

Intergovernmental Panel on Climate Change (IPCC). 2007. Climate Change 2007. Contribution of Working Group II to the Fourth Assessment Report of the IPCC: Impacts, Adaptation and Vulnerability. Tech. rep. Cambridge University Press.

Kane, S. and J.F. Shogren. 2000. Linking Adaptation and Mitigation in Climate Change Policy. *Climatic Change* 45 (1), 75–102.

Pittel, K. and D. Rübbelke. 2013. International Climate Finance and Its Influence on Fairness and Policy. *The World Economy* 36 (4), 419–436.

Zehaie, F. 2009. The Timing and Strategic Role of Self-Protection. *Environmental and Resource Economics* 44 (3), 337–350.

Appendix

A.1. General Solution of Nash Bargaining

$$B = [TC^f - s - TC^N][tc^f + s - tc^N] \rightarrow \max_{f^a, f^d, f^m} \tag{A.1}$$

$$\frac{\partial B}{\partial f^\bullet} = \frac{\partial TC^f}{\partial f^\bullet}\left[tc^f + s - tc^N\right] + \frac{\partial tc^f}{\partial f^\bullet}\left[TC^f - s - TC^N\right] = 0 \tag{A.2}$$

$$-\frac{\frac{\partial TC^f}{\partial f^\bullet}}{\frac{\partial tc^f}{\partial f^\bullet}} = \frac{[TC^f - s - TC^N]}{[tc^f + s - tc^N]} \tag{A.3}$$

Optimal $f^\bullet$ when impact on aggregate costs is zero, i.e., marginal global benefits of funding equal marginal costs of funding.

Chapter 10

Funding International Climate Protection: How to Overcome the Bergstrom Paradox?

Wolfgang Buchholz

University of Regensburg and CESifo Munich,
Universitätstrasse 31, 93053 Regensburg, Germany,
wolfgang.buchholz@wiwi.uni-regensburg.de

Richard Cornes

Australian National University,
Canberra, Australia

Wolfgang Peters

European University Viadrina,
Große Scharrnstrasse 59,
15230 Frankfurt (Oder), Germany
peters@europa-uni.de

Dirk Rübbelke

Technische Universität Bergakademie Freiberg
Schloßplatz 1, 09599 Freiberg, Germany
dirk.ruebbelke@vwl.tu-freiberg.de

*Very sadly, in the final stage of the preparation of this paper, our friend and esteemed colleague Richard Cornes passed away. We gratefully acknowledge financial support from the German Federal Ministry of Education and Research, FKZ 01LA1139A.

Abstract
In a world where the public good "climate protection" is provided via voluntary contributions, a country may introduce international transfers unilaterally in order to induce other countries to raise their contributions to this public good. With it underprovision of climate protection and inefficiency may be reduced. However, it is not obvious whether both donor and recipient are better off with funding. As in its pure form, funding climate protection only favors the donor (cf. Bergstrom, 1989), an additional instrument needs to restrict donor's advantages. Such a modified mechanism can be understood as a compromise between two bargaining partners searching for a solution which splits their gains in a fair way.

1. Motivation

It is a basic result, first observed by Bergstrom [1989], Boadway *et al.* [1989], Ihori [1996] and in more detail by Boadway *et al.* [2007] that subsidization of voluntary public-good contributions may have quite unexpected and somehow counterintuitive effects in the sense that the recipients of a conditional transfer do not necessarily gain from that measure or even lose although they have been subsidized.[1] Bergstrom's argument crucially depends on the assumption that the agents involved have identical preferences and that the subsidy payments are financed by lump-sum taxes. In the following, we show that this paradoxical effect is more general. Like the existing literature on this topic, our analysis also refers to the neighboring context of the matching literature by Guttman [1978, 1987].[2] However, we focus on subsidization of monetary public-good contributions as e.g., donations to charities (where subsidies are given through tax deductions) or transfers between countries that incentivize physical contributions

[1] Guttman [1981] provided a first hint in that direction.

[2] For similar results in a bi- or multilateral context see e.g., Althammer and Buchholz [1993]. Furthermore, Buchholz *et al.* [2015] present a diagrammatic exposition of an analogue argument within an aggregative game approach according to Cornes and Hartley [2007]. For a depiction of applications of the aggregative game approach in the environmental context, see Cornes [2015].

to a public good as e.g., greenhouse gas (GHG) abatement which are made to combat climate change.

We concentrate on contribution mechanisms that can emerge and be sustainable without a central authority, and that, therefore, may be particularly relevant especially for the provision of international public goods. A prominent example is climate finance which will be implemented under the roof of the United Nations Framework Convention on Climate Change's (UNFCC) negotiation process. Although it was an outcome of COP 16 at Cancun 2010 where industrialized countries promised to support less developed countries, it stands for a unilateral measure of a group of countries which try to enforce a voluntary bottom-up process to overcome the stagnancy in the annual COP negotiations in the post-Kyoto phase.

Climate finance especially bases on pioneering industrialized countries which engage in motivating emerging or developing countries in their engagement for reducing GHG emissions.[3] Here, "rich" countries may be interested in incentivizing "poor" countries such that by using transfers the latter are encouraged to participate in a larger scale of climate protection. Under the UNFCCC roof, it is discussed whether these measures should be done in form of unconditional transfers or conditional subsidies. E.g., funding which is earmarked for reducing the developing countries' vulnerability, i.e., for reducing their potential damage costs may be understood as nearly unconditional as this kind of funding does not incentivize a change in recipient's behavior through a regulation of the relative price of GHG abatements. Otherwise, conditional transfers directly aim at a price adjustment that reinforces the demand for climate protection of recipient countries. We analyze both kinds of funding mechanisms and ask whether and under which additional condition the donor as well as the recipient countries may find it advantageous to install such a climate fund.

[3] As Pickering *et al.* [2015] argue, developed countries have relied heavily on aid budgets to fulfill their pledges to raise their funding of climate change activities in developing countries. This instance adds some complexitiy which — however — is beyond the scope of our analysis.

2. Conditional Transfers

There are two agents, a donor D and a recipient R country, who have benefited functions $D(x_d, G)$ and $R(x_r, G)$ where x_i denotes D's and R's private consumption and G is public-good supply, for example reduction of GHG emissions. Both have standard properties, i.e., they are strictly quasi-concave and strictly monotone increasing and at least twice continuously differentiable. Moreover, we suppose that both private and public consumption are strictly normal in demand. The initial endowment ("income") of donor and recipient are w_d and w_r, respectively. The public good is produced by a linear summation technology as for climate protection it does not matter where GHG emissions are reduced. Furthermore, the marginal rate of transformation between the public and the private good is normalized to one since income can either be used for private consumption or for making a public-good contribution g_d and g_r.[4]

As long as donor and recipient are not engaged in some kind of funding climate protection, both voluntarily contribute as "price taker" with identical prices. Furthermore, we assume initially an interior subscription equilibrium $(\overline{g}_d, \overline{g}_r) > 0$. This public-good game yields an inefficient allocation suffering from an underprovision of climate protection. Now the "rich" donor considers to overcome this situation through the introduction of unilateral transfers channelled towards the "poor country". The donor unilaterally subsidizes the recipient's GHG mitigation that exceeds the previous amount of recipient's contributions $\overline{g}_r$. These incremental contributions are subsidized by an amount $s \cdot [g_r - \overline{g}_r]$. There are instruments in environmental policy pursuing such an approach. The Global Environment Facility of the World Bank, for example, subsidizes only incremental activities in developing countries while money stems from developed countries.[5]

Assuming a constant funding rate s then makes recipient's flat contribution cheaper. Thus, recipient's consumption becomes

[4]Thus, the aggregate budget constraint which defines feasible allocations reads $G = g_d + g_r = (w_d - x_d) + (w_r - x_r)$.
[5]On a depiction of this mechanism, see e.g., Rübbelke [2011].

$x_r = w_r - g_r + s \cdot [g_r - \overline{g}_r]$. The respective consumption on donor's side considers own flat contribution and costs of funding such that $x_d = w_d - g_d - s \cdot [g_r - \overline{g}_r]$. This simple funding mechanism changes D's and R's effective payoff to

$$D(w_d - g_d - s \cdot [g_r - \overline{g}_r]; g_d + g_r) \quad \text{and} \quad R(w_r - g_r + s \cdot [g_r - \overline{g}_r]; g_d + g_r). \tag{1}$$

Consequently, the respective first-order conditions are[6]

$$D_G - D_x = 0 \quad \text{and} \quad R_G \cdot \frac{1}{1-s} - R_x = 0. \tag{2}$$

These two equations determine the non-cooperative Nash equilibrium which can be used to analyze the impact funding has on donor's and recipient's behavior. By the usage of some abbreviations N_j^i for country i's preference properties of normality in demand of good j,[7] we can easily check the comparative statics of the public-good game in the presence of funding. Totally differentiating the equilibrium conditions (2), we obtain a linear system that can be solved for

$$\begin{pmatrix} \frac{dg_d}{ds} \\ \\ \frac{dg_r}{ds} \end{pmatrix} = \begin{pmatrix} N_x^d + N_G^d & N_x^d + sN_G^d \\ \\ N_x^r & N_x^r + N_G^r(1-s) \end{pmatrix}^{-1} \times \begin{pmatrix} -[g_r - \overline{g}_r]\, N_G^d \\ \\ [g_r - \overline{g}_r]\, N_G^r + R_G \frac{1}{(1-s)^2} \end{pmatrix}. \tag{3}$$

Thus, we are now able to show the economic consequences of introducing or extending the incentives of funding.

Proposition 1. *Comparative statics of funding climate protection.*

---

[6]Subscripts denote partial derivatives.

[7]As normality in demand is a property of preferences, and in turn, the utility function, we define for the recipient $N_x^r \equiv R_{xG} - R_{GG} \cdot R_x/R_G$ and $N_G^r \equiv R_{xG} \cdot R_x/R_G - R_{xx}$, respectively; for the donor, we have $N_x^d \equiv D_{xG} - D_{GG} \cdot D_x/D_G$, $N_G^d \equiv D_{xG} \cdot D_x/D_G - D_{xx}$. As we assume strict normality of both goods, all these terms become strictly positive, cf. Chiang and Wainwright [2005: 379] or Buchholz, Cornes and Peters [2008].

Given an interior subscription equilibrium $(\overline{g}_d, \overline{g}_r) > 0$, *both agents' contribution to the public good show expected results for an increase in the subsidy rate s. While the recipient feels an incentive to extend the public-good provision* $dg_r/ds > 0$, *the donor is engaged in an offsetting behavior* $dg_d/ds < 0$. *Nevertheless, the net effect on public-good supply G is positive.*

Proof. According to equation (3), we have some comparative statics:

$$\frac{dg_d}{ds} = -\frac{[g_r - \overline{g}_r]}{1-s} - \frac{R_G\left[N_x^d + sN_G^d\right]\frac{1}{(1-s)^2}}{\det} < 0, \tag{4}$$

$$\frac{dg_r}{ds} = \frac{[g_r - \overline{g}_r]}{1-s} + \frac{R_G\left[N_x^d + N_G^d\right]\frac{1}{(1-s)^2}}{\det} > 0, \quad \text{and} \tag{5}$$

$$\frac{dG}{ds} = \frac{dg_d}{ds} + \frac{dg_r}{ds} = \frac{R_G N_G^d \frac{1}{1-s}}{\det} > 0. \tag{6}$$

The calculations make use of an abbreviation det which stands for the determinant of the matrix in (3). As det is always positive, the Nash equilibrium is stable and unique.[8] Furthermore, equation (6) results from rearranging terms and combining equations (4) and (5) with the determinant det. □

As funding was introduced to reduce inefficiency, we can state that at least the underprovision of climate protection declines. However, it is not obvious whether both agents are better off with this kind of funding. Although we expect the recipient to gain from donor's gift, it may be the donor who suffers from his own generosity. Amazingly, checking the consequences of an increased funding rate yields opposite consequences than those expected, and thus, Bergstrom's [1989] paradox.

Proposition 2. *The Bergstrom paradox.*

---

[8] As we have assumed a normal demand for both private and public goods, the determinant $\det = (1-s)[N_G^d \cdot N_x^r + N_x^d \cdot N_G^r + N_G^d \cdot N_G^r]$ is always positive. For stability and uniqueness cf. Tirole [1993: 324].

Assume an interior equilibrium with $(\overline{g}_d; \overline{g}_r) > 0$. *Then any increase in the funding rate s will favor the donor while penalizing the recipient.*

Proof. By considering the first-order conditions (2) and the comparative statics of Proposition 1, totally differentiating donor's and recipient's payoff with respect to s yields

$$\begin{aligned}\frac{dD}{ds} &= -D_x\left[g_r - \overline{g}_r\right] + \left[D_G - D_x s\right]\frac{dg_r}{ds} \\ &= \frac{D_x R_G\left[N_x^d + N_G^d\right]}{(1-s)\det} > 0, \quad \text{and} \qquad (7)\\ \frac{dR}{ds} &= R_x\left[g_r - \overline{g}_r\right] + R_G\frac{dg_d}{ds} \\ &= -\left(\frac{R_G}{1-s}\right)^2\frac{N_x^d + sN_G^d}{\det} < 0. \qquad (8)\end{aligned}$$

In general, both agent's benefit changes in opposite direction. □

Bergstrom's paradox says that, although the recipient is subsidized through the donor, he is unable to gain from that gift. In equilibrium, recipients realize a disadvantage of lower prices for the public good (i.e., contributions net of subsidies).[9] If such "relative prices" can be manipulated strategically by one of the parties (here through a subsidy on climate protection), they may gain at the expense of their opponents. The incentive structure of the funding mechanism is so strong that the recipient is solely responsible for the increase in public-good supply ($dG/ds > 0$), while the donor country decreases its own countributions ($dg_d/ds < 0$). And even when the donor calculates the total financial burden, own measures for climate protection plus transfers to support the recipient, he pays, all in all, not more but even less than before; i.e., donor's private consumption slightly

[9]This phenomenon was discussed in a similar context as "advantage of being poor" by Konrad [1994] or "advantage of being less skilled" by Buchholz *et al.* [1998].

increases ($dx_d/ds = -d(g_d + s[g_r - \overline{g}_r])/ds = R_G N_x^d/(1-s)\det > 0$). Consequently, subsidizing climate protection via funding incentivizes the recipient so strong that the donor favors from two effects: (i) total public-good supply is increased, and (ii) funding yields a more than 100% crowding out such that the "supposed donor" actually becomes in fact an "in-kind transfer recipient". Although the donor country bears the financial burden of subsidizing recipient's additional abatement activities, this country cuts its own GHG abatements by more than 100% of these costs such that donor's private consumption increases with funding climate protection. Or, in terms of donor's virtual income ($I_d = w_d + g_r - s[g_r - \overline{g}_r]$), which consists of its initial endowment plus recipient's gross abatement cost minus funding expenses, donor's disposible income increases through the unilateral transfer mechanism. And, therefore, donor's demand for private consumption is enlarged. Thus, he favors twice because his private as well as public consumption are extented.

Summarizing, funding reduces the inefficiency (underprovision) in terms of GHG abatements but does not yield a Pareto improvement. The donor gains at the expense of his opponent such that the recipient is advised not to accept a gift through a conditional funding. Consequently, the considered funding scheme will not work when both implicitly contracting parties would have to accept the rules of this mechanism.

3. Avoiding the Bergstrom Paradox

In the previous section, we have learnt that donor's offsetting behavior is the main origin for the recipient not to agree to such a mechanism. If the donor, however, tries to overcome the inefficiency of voluntary contributions through the usage of a funding incentive, the mechanism needs to be modified such that it is promising for both parties. Thus, the donor has to make some concessions: he either promises an additional, but unconditional transfer or not to offset (i.e., donor's contributions remain unchanged at the equilibrium level without any funding $\overline{g}_d$).

3.1. *First attempt: Unconditional transfers*

As with a pure funding mechanism, the recipient does not gain. An additional income transfer could be seen as an appropriate measure to overcome these negative consequences. Such a transfer aims at improving the recipient's ability to pay for climate protection as well as private consumption. Thereby, the unconditional transfer is more balanced from recipient's point of view. In that case, an unconditional (lump-sum) transfer from donor to recipient may yield a Pareto improvement. Thus, we introduce a transfer t, which obviously changes the initial "net income" of both parties such that D's and R's effective payoff become

$$\begin{aligned} &D(w_d - t - g_d - s[g_r - \overline{g}_r]; g_d + g_r), \quad \text{and} \\ &R(w_r + t - g_r + s(g_r - \overline{g}_r); g_d + g_r), \text{ respectively.} \end{aligned} \tag{9}$$

Consequently, the respective first-order conditions remain unchanged. However, the comparative statics with respect to t show whether the transfer works in the right direction. Totally differentiating the equilibrium conditions (2), we obtain a linear system that can be solved for

$$\begin{pmatrix} \frac{dg_d}{dt} \\ \\ \frac{dg_r}{dt} \end{pmatrix} = \begin{pmatrix} N_x^d + N_G^d & N_x^d + sN_G^d \\ \\ N_x^r & N_x^r + N_G^r(1-s) \end{pmatrix}^{-1} \begin{pmatrix} -N_G^d \\ \\ N_G^r \end{pmatrix}. \tag{10}$$

Thus, we are now able to show the economic consequences of such a transfer.

Proposition 3. *Comparative statics of an unconditional transfer (extended Warr neutrality).*

Given an interior subscription equilibrium $(\overline{g}_d, \overline{g}_r) > 0$, *both agents' contribution to the public good show expected results for an increase in a lump-sum transfer* t. *While the recipient feels an incentive to extend the public-good provision* $dg_r/dt > 0$, *the donor is engaged in an offsetting behavior* $dg_d/dt < 0$. *Nevertheless, the net effect on public-good supply* G *is nil. Thus, we have an extended Warr* [1983] *neutrality.*

Proof. According to equation (10) and by inserting the definition for det, we have some comparative statics:

$$\frac{dg_d}{dt} = -\frac{1}{1-s} < 0, \tag{11}$$

$$\frac{dg_r}{dt} = \frac{1}{1-s} > 0, \tag{12}$$

$$\frac{dG}{dt} = \frac{dg_d}{dt} + \frac{dg_r}{dt} = 0, \quad \text{and} \tag{13}$$

$$\frac{dx_r}{dt} = 1 - (1-s)\frac{dg_r}{dt} = 0. \tag{14}$$

Consequently, as x_r and G remain unchanged, it follows from the aggregate budget constraint that consumption of the donor x_d does not alter too. □

As the transfer was introduced to overcome recipient's losses from funding, we can state that as long as both parties initially take positive measures to combat climate change, an unconditional transfer remains without any allocative consequences. Therefore, this first attempt fails its objective, i.e., funding accompanied by a lump-sum transfer does not yield a Pareto improvement.

3.2. *Second attempt: Commitment not to offset*

Obviously, the origin of Bergstrom's paradox is caused by donor's offsetting behavior. Thus, to overcome that problem a commitment of the donor not to change the contribution to international climate protection seems to be appropriate. Under such a modification, it is only the recipient who makes an active choice, and recipient's new first-order condition reads

$$R_G(w_r - g_r + s[g_r - \overline{g}_r]; \overline{g}_d + g_r) \cdot \frac{1}{1-s} - R_x(w_r - g_r + s[g_r - \overline{g}_r]; \overline{g}_d + g_r) = 0; \tag{15}$$

which, in turn, yields an altered incentive for the recipient to engage in climate protection. Climate funding now yields additional mitigation activities of

$$\frac{dg_r}{ds} = \frac{[g_r - \overline{g}_r]\, N_G^r + R_G \frac{1}{(1-s)^2}}{N_x^r + (1-s)N_G^r} > 0. \tag{16}$$

Proposition 4. *Overcoming the Bergstrom paradox.*

The modified funding mechanism where the donor pledges not to offset recipient's increased efforts yields a Pareto improvement, at least for a sufficiently small subsidy rate s.

Proof. Totally differentiating donor's and recipient's utility yields

$$\frac{dD}{ds} = [D_G - D_x s]\frac{dg_r}{ds} - D_x[g_r - \overline{g}_r], \quad \text{and} \tag{17}$$

$$\frac{dR}{ds} = \underbrace{[R_G - (1-s)R_x]}_{=0}\frac{dg_r}{ds} + R_x \cdot [g_r - \overline{g}_r] = R_x \cdot [g_r - \overline{g}_r] > 0. \tag{18}$$

While the latter equation shows that under the modified incentive, the recipient always gains and donor's benefit does only improve for small funding rates s. As $\lim_{s \to 0}[g_r - \overline{g}_r] = 0$, there exists an $\varepsilon > 0$ such that for all $0 < s < \varepsilon$ donor's as well as recipient's benefits change in the right direction. □

This guarantees that both the donor and the recipient become better off by the pledge of the donor not to offset the increased effort of the recipient.

3.3. *Importance of subsidizing incremental GHG abatements*

Equation (17) directly shows how essential it is that the donor only subsidizes the GHG mitigation exceeding recipient's previous contribution $\overline{g}_r$. In case of subsidizing total flat contributions g_r, the donor does not necessarily gain from funding even for small subsidy rates

$$\left.\frac{dD}{ds}\right|_{\overline{g}_r=0} = [D_G - D_x s]\frac{dg_r}{ds} - D_x g_r. \tag{19}$$

A similar argument as in case of subsidizing incremental GHG abatements that guarantees a general improvement for the donor is not possible. It turns out that we need supplementary conditions to make funding profitable for the donor.

Proposition 5. *Funding total flat contributions.*

For a small subsidy rate, the donor becomes only better off if the elasticity of recipient's contribution with respect to donor's funding $\eta := s/g_r \cdot dg_r/ds$ *exceeds unity, i.e., the recipient responds elastically to donor's incentive.*

Proof. For $s \to 0$, it turns out that $g_d \to \overline{g}_d$, and therefore $D_G \to D_x$ such that $dD/ds > 0$ is equivalent to $\eta > 1$. □

However, this supplementary condition which yields a Pareto improvement is only fulfilled for specific preferences as well as distributions of initial wealth. Therefore, to avoid that funding may fail, it seems to be appropriate to shorten the incentive of funding to the change in recipient's behavior relative to GHG abatements before the introduction of funding. A good example for this purpose is the Global Environmental Facility of the World Bank which subsidizes only incremental measures of developing countries. As we have shown, such a mechanism that incentivizes changes in behavior only and complements this with donor's commitment of a refrain from offsetting yields a Pareto improvement which is independent of assumptions on preferences and the distribution of income between industrialized and developing countries.

4. Efficient Choice of Funding

In case of funding accompanied by a commitment of the donor country not to cut back its own abatement policy, we have an effective measure to overcome the inefficiency of a non-cooperative climate policy. This implicit cooperation asks for an efficient use of the instrument. On the one hand, we can look for a reasonable policy for donor's point of view or, on the other side an optimal compromise for both parties, which can be generated by Nash bargaining or similar concepts on negotiations, as they may be interesting.

4.1. *Donor's optimal funding rate*

As the donor benefits as long as he subsidizes the recipient with moderate funding rates s, the question arises on which rate a donor country should offer its opponent such that the payoff becomes optimal. Hence, equation (17) can be used as a first-order condition. By solving for an optimal subsidy rate, we obtain

$$s^* = \frac{D_G}{D_x}\frac{\mu}{1+\mu}, \tag{20}$$

where $\mu := s \cdot d[g_r - \overline{g}_r]/ds \cdot [g_r - \overline{g}_r]$ refers to the elasticity of recipient's incremental contribution with respect to donor's funding activity. Obviously, the optimal rate s^* increases with the success of this measure when incentivizing the recipient; i.e., with a rising μ. Additionally, this rate builds a lower bound for a compromise that parties may be interested in. Up to s^*, both countries become better off and the recipient will be favored even by higher rates. Therefore, Pareto improvements end with s^*.

4.2. *Negotiating on the extent of funding*

As both parties, donor and recipient, overcome the inefficient non-cooperative equilibrium through applying and accepting the conditions of climate funding, for a next step it seems to be appropriate to look for a cooperative agreement on which rate of funding to choose. Starting from the non-cooperative benchmark of the subscription equilibrium without any funding, the introduction of a conditional subsidy is advantageous for both even for $s > s^*$.

Figure 1 shows which benefit combinations donor and recipient can achieve with funding. Point A builds the benchmark point with $s = 0$; up to point B both parties gain from subsidies $s \in (0; s^*]$.

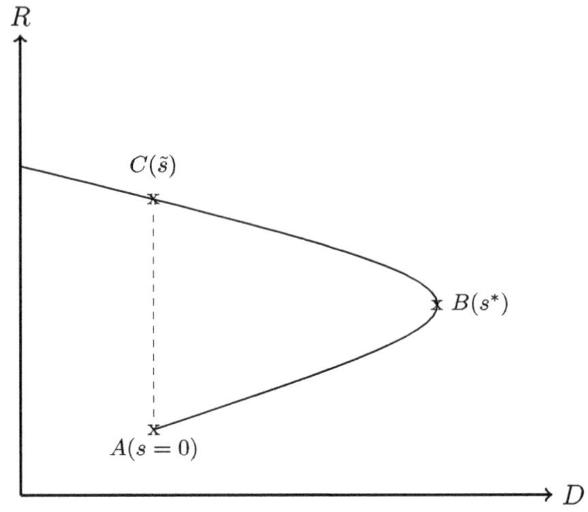

Figure 1. Benefit Combinations for Different Funding Rates.

Until a subsidy rate $\tilde{s} > s^*$, the recipient gains from funding while the donor is still better off than in the fall-back position A. Thus, when both parties negotiate (for example through Nash bargaining) with the outside option A, they agree on a compromise between B and C. This does, however, presuppose that both accept the fall-back position A. Whether the donor agrees on such a procedure can be doubted as it is the donor who unilaterally installs a Pareto improvement through introducing a funding system. This unilateral action yields a Pareto improvement and even the recipient agrees with the rules of the mechanism. Therefore, point B becomes the starting point for negotiations, and thus there is no scope for further trials anymore. Consequently, we have two different outcomes of an agreement between both parties. Either they agree on a point somewhere between B and C (Nash bargaining right from the non-cooperative start), or B as a process of step by step negotiations only accepting Pareto improvements.

If both parties anticipate subsequent negotiations and the respective outcomes, the recipient country is advised not to participate in a step-by-step procedure of Pareto improvements as it obstructs a successful bargaining, which ends up between B and C. The donor country should insist on its unilateral measure, ignoring any proposal to negotiate on a compromise. A "take-it-or-leave-it" offer seems to be appropriate.

5. Conclusion

For a successful application, climate funding has to be accepted with unanimity by donor and recipient. In its pure form, funding via one-sided matching only favors the donor, due to the induced additional room for maneuver concerning free-riding. This effect is largely disregarded in the applied literature assessing the effects of international climate transfers, possibly because it tends to be overlain by other effects, e.g., the beneficial exploitation of international cost differentials in greenhouse gas abatement. In order to address the considered adverse influence of conditional transfers, an additional instrument has to restrict the donor's respective extension of his room of maneuver.

While the ban of offsetting behavior is strong enough to favor the recipient, its incentive to enforce additional contributions of the recipient suffers from a countervailing income effect. However, at least for small subsidies the income effect vanishes. This modified mechanism can be understood as a compromise between two bargaining partners searching for a solution which splits its gain in a fair way. Depending on the starting point of negotiations, we end up in Nash bargaining or an optimal subsidy from the donor's perspective. It is a matter of negotiation power which starting point seems to be appropriate. On the one hand, a unilateral policy can hardly be prohibited as it is in line with both parties' interests. On the other hand, if the recipient does not accept a step-by-step procedure this hard-line behavior may yield Nash bargaining as a compromise. Therefore, the proposal of a step-by-step policy is a slow poison for recipient countries. "Wait and see" seems to be more advantageous for developing or emerging countries. They can gain more through tough negotiations.

References

Althammer, W. and W. Buchholz. 1993. Lindahl Equilibria as the Outcome of a Non-Cooperative Game. *European Journal of Political Economy* 9, 399–405.

Bergstrom, T. 1989. Love and Spaghetti, the Opportunity Cost of Virtue. *Journal of Economic Perspectives* 3, 165–173.

Boadway, R., P. Pestieau and D. Wildasin. 1989. Tax-transfer Policies and the Voluntary Provision of Public Goods. *Journal of Public Economics* 39, 157–176.

Boadway, R., Z. Song and F. J. Trembley. 2007. Commitment and Matching Contributions to Public Goods. *Journal of Public Economics* 91, 1664–1683.

Buchholz, W., L. Nett and W. Peters. 1998. The Strategic Advantage of Being Less Skilled. *Economics Letters* 60, 35–39.

Buchholz, W., R. C. Cornes, W. Peters and D. Rübbelke. 2015. Pareto Improvement through Unilateral Matching of Public Good Contributions: The Role of Commitment. *Economics Letters* 132, 9–12.

Buchholz, W., R. C. Cornes and W. Peters. 2008. Existence Uniqueness and Some Comparative Statics for Ratio and Lindahl Equilibria. *Journal of Economics* 95, 167–177.

Chiang, A. C. and K. Wainwright. 2005. Fundamental Methods of Mathematical Economics, 5th edn. Boston, MA: McGraw-Hill.

Cornes, R. C. 2015. Aggregative Environmental Games. *Environmental and Resource Economics*, online first.

Cornes, R. C. and R. Hartley. 2007. Aggregative Public Good Games. *Journal of Public Economic Theory* 9, 201–209.

Guttman, J. M. 1978. Understanding Collective Action: Matching Behavior. *American Economic Review* 68, 251–255.

Guttman, J. M. 1981. Matching Behavior and Collective Action: Theory and Experiments. Discussion paper. Hebrew University of Jerusalem.

Guttman, J. M. 1987. A Non-Cournot Model of Voluntary Collective Action. *Economica* 54, 1–19.

Ihori, T. 1996. International Public Goods and Contribution Productivity Differentials. *Journal of Public Economics* 61, 139–154.

Konrad, K. A. 1994. The Strategic Advantage of Being Poor: Private and Public Provision of Public Goods. *Economica* 61, 79–92.

Pickering, J., J. Skovgaard, S. Kim, J. T. Roberts, D. Rossati, M. Stadelmann, and H. Reich. 2015. Acting on Climate Finance Pledges: Inter-Agency Dynamics and Relationships with Aid in Contributor States. *World Development* 68, 149–162.

Rübbelke, D. T. G. 2011. International Support of Climate Change Policies in Developing Countries: Strategic, Moral and Fairness Aspects. *Ecological Economics* 70, 1470–1480.

Tirole, J. 1993. The Theory of Industrial Organization, 6th edn. Cambridge, MA and London, UK: MIT Press.

Warr, P. G. 1983. The Provision of a Public Good Is Independent of the Distribution of Income. *Economics Letters* 13, 207–211.

Chapter 11

Conditional Transfers in International Climate Policy — The Case of Adaptation Transfers

Karen Pittel

Ifo Institute — Leibniz Institute for Economic Research at the University of Munich, Poschingerstrasse 5, 81679 Munich, Germany
pittel@ifo.de

Dirk Rübbelke

Technische Universität Bergakademie Freiberg, Chair for Economics Especially Resource Economics, Schloßplatz 1, 09599 Freiberg, Germany
Dirk.Ruebbelke@vwl.tu-freiberg.de

Abstract

In order to overcome the under-provision of global public goods, various policy approaches have been proposed. In the climate policy arena, international transfers are frequently seen as an effective means to raise the provision of the global public good "climate change mitigation". This chapter focuses on a specific type of international transfer that aims at raising mitigation while also reducing the damages from climate change: conditional adaptation support.

Especially since the Conference of the Parties (COP) in Copenhagen in 2009, preparations are ongoing to significantly expand international transfers that help developing countries to adapt to climate change. While there are extensive discussions in the policy arena about the required amount of adaptation funding and the best ways to raise, manage and disburse these funds, hardly any attention is paid to the international allocative effects of these transfers. The answer to the question of "why" international adaptation transfers are paid at all, is often relegated to fairness considerations only. As adaptation benefits are largely local and adaptation transfers reduce the recipients' incentives to contribute to climate change mitigation, one would, however, expect at least unease in donor countries about plans to significantly expand international adaptation support on these grounds.

In this study, we compare two alternative conditional transfer schemes: one plainly subsidizes mitigation efforts, while the other provides adaptation support which is conditional on other agents' mitigation contributions. Disregarding distributional and fairness aspects, the chapter evaluates and compares the allocative effects of either policy scheme. It is shown that while both policy schemes can be beneficial for developing as well as industrialized countries, this outcome relies strongly on the productivity of mitigation and adaptation technologies.

1. Introduction

One of the largest obstacles to the efficient provision of global public goods is the lack of an international coercive authority that can enforce an efficient provision level. So, global public goods like climate change mitigation rely solely on voluntarily contributions and measures of individual countries. A logical prerequisite for a country to voluntarily engage in such activities would be the expectation to be better off as a result.

International support for adaptation in other countries seems to be a surprising phenomenon in this context. Climate change adaptation creates mostly local benefits and hardly any international spillovers that mitigation efforts of other countries would induce. So why should a country choose to delegate funds to international adaptation support? In this chapter, we take a look at this phenomenon of adaptation transfers and investigate whether and under which conditions introducing adaptation support conditionality might not only bring about Pareto improved allocations but might even make adaptation support more attractive from the donor's point of view than direct mitigation support.

Efforts to internationally coordinate climate change policies have been largely delegated to conferences, so-called Conferences of Parties (COPs), under the United Nations Framework Convention on Climate Change. In recent climate negotiations, international support of developing countries in adapting to climate change has become a strongly discussed issue.[1] An important point of controversy is the share of international climate-related funding that should be dedicated to adaptation support. The Cancun Accords, agreed internationally at the COP in Cancun in 2010, stipulate that adaptation "must be addressed with the same priority as mitigation" [UNFCCC, 2011]. Thus, it is conceivable to expect that a large share of the USD 100 billion, having been pledged by industrialized countries in Cancun as an annual climate-related transfer towards the developing world by 2020, will be dedicated to adaptation tasks.

From an allocational perspective, it might be a bit surprising at a first glance that such a large amount of money will be transferred internationally for adaptation projects in the developing world when the donor countries hardly benefit directly from these projects. Benefits of adaptation projects are largely local and thus, "[i]n contrast to mitigation, the benefits of adaptation are excludable" [Barrett, 2008: 254] from an individual country's point of view.[2] Therefore, Tol [2005: 577] points out: "[n]ational governments and international organizations have little to do with adaptation, and should not try."[3]

[1]Before, as Paavola and Adger [2006: 595] remark, there has been a "longstanding unease in the policy community with regard to adaptation" and this "originates from fears that the acknowledgement of a possibility of adaptation could distract international efforts to mitigate climate change".

[2]Kane and Shogren [2000: 86] stress: "Adaptation is a private good in which the benefits of reduced severity" [of climate change effects] "accrue only to the nation".

[3]Of course, adaptation can have other effects on development that might provide some additional incentives for adaptation support (political stability, preventing migration, distributional concerns, to name a few). As we are, however, interested in analyzing whether there might be a case for adaptation support from the point

Beyond, there are also strategic reasons for unease with regard to adaptation support on the international level, because such support tends to reduce transfer-receiving countries' incentives to contribute to the global public good "mitigation". This is because adaptation reduces the vulnerability of these countries and hence also tends to decline their mitigation benefits. Put differently, to some extent mitigation and adaptation tend to be substitutes. As the merits of the global public good "mitigation" are globally non-excludable, a decline in developing countries' contributions to this global public good — in turn — harms industrialized countries. Thus instead of awarding adaptation assistance, it seems more beneficial for donor countries to provide international support of mitigation only.

In order to find out whether there are nevertheless good reasons from an allocation perspective to provide international adaptation support, we compare the effects of different transfer schemes and consider whether these schemes can induce a Pareto improvement. This can help to assess whether the expectation that "some Parties choose to pursue voluntary cooperation in the implementation of their nationally determined contributions to allow for higher ambition in their mitigation and adaptation actions" [Paris Agreement, UNFCCC 2015] can be seen as realistic.

As e.g., Buchholz and Konrad [1995] and Cornes and Hartley [2007] demonstrate, even unconditional international transfers can, under specific conditions, induce a Pareto improvement. Yet, because unconditional transfers only generate income effects, they are associated with dispersion losses as transfer receptors do not employ the whole transfer for climate change mitigation but use a portion for raising private good consumption (this share of private good consumption can be up to one, see e.g., Ono [1998] who considers the possibility of corner solutions). The augmentation of developing countries' private good consumption itself does, however, not improve the welfare of donor countries.

---

of view of industrial countries even in the absence of these issues, we do not integrate them into our analysis in this chapter.

In contrast, transfers provided conditional on a rise in developing countries' mitigation efforts are more effective in increasing the level of the global public good "mitigation" because they induce — additional to income effects — price effects, i.e., they change the effective price of mitigation. In the case of subsidies for mitigation conducted by developing countries, for example, the effective price of mitigation falls, leading to substitution between the provision of the public good mitigation and the private good. Consequently, conditional transfers are more desirable from the donor countries' point of view. The transfer-induced increase in global public good contribution by the developing world yields — via its positive global externalities — a return on the donor countries' investments in international transfers. As observed by Heuson *et al.* [2012: 5], "[d]ue to its public-good character, mitigation is the only channel through which the industrialised country can profit in this respect — contrary to the private good of adaptation".[4]

In our model, which is based on an approach developed by Ihori [1996] and which can be classified as a contribution to the theory of voluntary public good provision,[5] conditional transfers are analyzed. We distinguish between conditional transfers provided in monetary terms (subsidies) and in-kind transfers. These in-kind transfers take the shape of support to climate change adaptation. We investigate induced changes in mitigation levels in developing countries as well as the conditions under which in-kind "adaptation" transfers are more beneficial, from both the industrialized and developing countries'

[4]As Buob and Stephan [2011a] state, the industrialized world can only benefit from investing in adaptation funding if it induces the developing world to raise its mitigation efforts, "which is the case if there is strict complementarity between mitigation and adaptation". In contrast to Buob and Stephan [2011a], in our setting adaptation funding can induce a rise in developing countries' mitigation although adaptation and mitigation are substitutes.

[5]Besides models on the voluntary provision of public goods, there is a second strand of literature which deals with the interplay between mitigation, adaptation and transfers. This is the Integrated Assessment Models (IAMs) literature, where the new generation of IAMs treats adaptation as endogenous and allows for policy analysis; see e.g., de Bruin *et al.* [2009] and Bosello *et al.* [2010].

point of view, than purely monetary mitigation subsidies. If we can determine conditions under which international adaptation support is more advantageous from all involved agents' perspective, then the current strong focus on adaptation transfers in the international climate policy arena might be justified from an allocative perspective.[6]

In order to investigate these aspects and to conduct a comparative statics analysis of potential influence factors, we transform standard welfare maximization problems into expenditure minimization problems (see e.g., Diamond and McFadden [1974] and Cornes [1992] for this dual approach). In our study, similar to the earlier analysis in a two-agent framework by Ebert and Welsch [2012], technological productivities play an important role. Yet, in contrast to Ebert and Welsch [2012], our study focuses on the influence of technological productivities on the effects of international transfers.

In detail we proceed as follows. Section 2 gives a short outline of the framework of this analysis. In Section 3, we concentrate on welfare maximization of the developing region within three scenarios. The first scenario describes the reference case where no transfers are provided and which is in line with the standard Nash case. Thereafter, we consider two scenarios where international mitigation subsidies are provided and where mitigation efforts in developing countries are matched by international adaptation transfers. We depict the influence of the respective policies on the optimization problem of the developing region and ascertain the modification of effective mitigation prices. In Section 4, we reformulate the welfare maximization problems of the developing and industrialized regions as expenditure minimization problems in order to analyze the regional welfare effects of introducing subsidies and matching transfers. In order to do so, we distinguish in a comparative

[6]Although no explicit conditionality of adaptation transfers is proclaimed in recent international agreements, there might be such a conditionality pending. Still, the rules for the future annual USD 100 billion transfers are not stipulated and the rules might be aligned with the developing countries' willingness to contribute to mitigation.

static analysis between different cases and scenarios. Section 5 concludes.

2. The Model

Consider a world consisting of a developing region D, and an industrialized region I. Both regions derive utility from the consumption of a private good bundle, y_i, $i = I, D$, and a global public good "world-wide mitigation efforts", $X = \sum_i x_i$, with x_i denoting the mitigation effort of region i. The benefits that a region derives from both the private good bundle and the public good depend on the level of adaptation a_i to climate change.

Firstly, adaptation increases the benefits obtained from private good expenses. Total private consumption $c_i = c_i(y_i, a_i)$ is the effective private consumption derived from expenses for the private good bundle y_i and adaptation a_i. The more adaptation takes place, the better the private good bundle (and its use) is protected against climate change, i.e., the higher is c_i (with $c_{iy_i} > 0$ and $c_{ia_i} > 0$). For the intuition behind this assumption, consider the following example: the benefits from owning a car will be higher if roads are not continuously flooded, i.e., the adaptation measure "building a dike" raises the benefits from private goods.

Secondly, the effectively consumable amount of mitigation benefits in region i depends on adaptation, $X_i^c = X_i^c(X, a_i)$ where X_i^c is the adaptation adjusted benefit from global mitigation X. The term "effectively consumable" is used to differentiate between mitigation efforts X and the benefits resulting from these mitigation efforts X_i^c. The latter depend on the level of adaptation in region i. Similarly to Auerswald *et al.* [2011: 4] who assume adaptation to be a substitute measure for self-insurance against the risk from climate change, we suppose that higher regional adaptation reduces the current (and future) costs from climate change and thus the benefits from mitigation in this region ($X_{ia_i}^c < 0$). Also Ingham *et al.* [2007: 5358] consider adaptation and mitigation to be substitutes, as "they are two different ways of reducing climate change damages, and if, say, the costs of adaptation fell, we expect society to do more adaptation and less mitigation". And Ingham *et al.* [2005a,b] show that adaptation and mitigation are in general substitutes, but they also identify cases where complementarity may prevail. For example, complementarity could arise when adaptation costs depend on the amount of mitigation conducted or, in an intertemporal context, if adaptation increases the incentives for mitigation by raising the value of future assets.[7] Buob and Stephan [2011b] investigate both the case where marginal costs of adaptation inversely depend on global mitigation and the case where marginal costs of adaptation are independent of global mitigation. Also Pittel and Rübbelke [2015] consider the case of substitutes and complements in a framework very similar to ours. They show that the welfare effects of climate policy might depend on adaptation and mitigation being substitutes or complements.

Please note that while we assume that adaptation in region i exerts a negative effect on the mitigation benefits in the same region, it seems reasonable to assume that the mitigation benefits in region j are unaffected by region i's adaptation efforts ($i \neq j$). Hence, as the adverse mitigation productivity effect of adaptation transfers is only experienced locally, it can be interpreted as a private effect (from the respective region's point of view).

Keeping the above defined relationships in mind, a region's welfare function is given by $U_i = U_i(c_i(y_i, a_i), X_i^c(X, a_i))$, i.e., the utility depends on the effectively consumable outputs or characteristics c_i and X_i^c of the goods y_i and $X = x_i + x_j$, respectively. Thus, in line with Lancaster [1966: 133], we break away from the "traditional" approach "that goods are the direct objects of utility" and suppose that it is instead the characteristics or outputs of the goods from which utility is derived; also see Gorman [1980] and Cornes and Sandler [1984] for such distinctions between goods and characteristics. Throughout, we assume that the characteristics behave like normal "goods".

In this general set-up, each region allocates its exogenously given income I_i to private goods, mitigation efforts and adaptation. To keep

[7] Also maladaptation may bring about complementarity of mitigation and adaptation [Ingham *et al.*, 2005a: 14].

the analysis tractable, we will apply some further specifications and simplifying assumptions in the following sections of the chapter. We will assume that the industrialized region provides conditional adaptation or monetary transfers towards the developing world. As our focus is on the welfare effects of international transfers channelled towards the developing region, we abstract from industrialized region's investment in domestic adaptation (i.e., $a_I = 0$).[8] Regarding the developing region, we assume that a_D is completely initiated by transfers from the industrialized region. One may argue that — due to better technological knowledge in industrialized countries — adaptation is more effective and cheaper when it is provided via transfers by industrialized countries and thus developing countries refrain from own "less effective" domestic adaptation efforts.

Put it differently, we focus on the situation in which

- industrialized countries can commit within the framework of an international agreement to channel conditional transfers towards the developing world,
- the developing region is too poor and/or inexperienced to invest in adaptation itself,[9] and
- the industrialized region can choose between two policies: subsidizing mitigation and financing adaptation in the developing region.

3. Welfare Maximization in Different Support Schemes

In this section, we distinguish between the three aforementioned scenarios: (1) no international transfer policy, (2) mitigation activities

[8] To focus exclusively on climate policy related flows of goods and payments across region's, we abstract from international trade in goods. For an analysis of the effects of international trade and terms-of-trade effects in the context of climate change, see, e.g., Schenker [2013] and, for an empirical analysis in the context of adaptation, Schenker and Stephan [2012].

[9] In reality, there is of course locally financed and conducted adaptation also in developing countries. We abstract from this type of adaptation in our model as it would induce further income and substitution effects which would blur the transmission channels of the conditional transfers that we focus on in this chapter.

in the developing region are subsidized by the industrialized world, and (3) the industrialized region grants adaptation support for the developing region conditional on its level of mitigation activity. By determining the conditions for the developing region's welfare maximization and ascertaining the resulting effective mitigation prices, we can show that each of these scenarios brings about different mitigation levels in the developing world. Based on these findings, we analyse welfare implications of the different international transfer schemes and varying technological productivities in Section 4 of this chapter.

3.1. *No-policy scenario*

If no transfer policy is conducted by the industrialized region, developing region D maximizes its welfare function (1) subject to its budget constraint (2a)[10]:

$$\max_{y_D, x_D} U_D(c_D(y_D, a_D), X_D^c(X, a_D)) \tag{1}$$

$$\text{s.t. } I_D = y_D + x_D. \tag{2a}$$

As adaptation is not financed by the industrialized region in this scenario, $a_D = 0$ holds and we assume for simplicity $c_D = y_D$ and $X_D^c = X$ in this case. Prices of both goods are normalized to unity. The first-order conditions for the optimization problem in (1) and (2a) give rise to the well-known equalization of the marginal rate of substitution between the two goods to their price ratio:

$$\frac{\frac{\partial U_D}{\partial X_D^c}}{\frac{\partial U_D}{\partial c_D}} = 1 \tag{3}$$

with $\frac{\partial c_D}{\partial y_D} = \frac{\partial X_D^c}{\partial x_D} = 1$. Policies that cause a decline in the right-hand side (RHS) and/or an increase in the left-hand side (LHS) of (3)

---

[10]Please note that the corresponding maximization problem of the industrialized region would look slightly different as there is no adaptation in the industrialized region. Thus, its welfare function can simply be stated as $U_I(y_I, X)$. The budget constraint is $I_I = y_I + x_I + T(x_D)$, where T are the funds transferred towards the developing region depending on the level x_D of its mitigation efforts.

would induce a higher level of region D's voluntary contributions to the global public good.

3.2. *Subsidies: Supporting mitigation through monetary transfers*

If the industrialized region subsidizes mitigation efforts in the developing region,[11] the modified budget constraint of the developing region reads

$$I_D = y_D + (1 - s)x_D, \quad 0 < s < 1 \tag{2b}$$

where s is the subsidy rate paid by the industrialized region for climate change mitigation efforts in the developing region. As adaptation is not financed by the industrialized region in this policy scenario, $a_D = 0$ and thus $c_D = y_D$ and $X_D^c = X$ still hold. Maximization of (1) subject to (2b) yields

$$\frac{\frac{\partial U_D}{\partial X_D^c}}{\frac{\partial U_D}{\partial c_D}} = 1 - s. \tag{4}$$

Compared with the no-policy scenario, the price ratio between private and public goods is lower and thus the mitigation level is higher in the subsidy case. The RHS of (4) can be interpreted as the effective price $p_D^{e,s}$ of public good provision and obviously this effective price has declined due to the international subsidy payments.[12]

3.3. *Matching: Adaptation support*

Instead of directly subsidizing mitigation, the industrialized region can finance in-kind transfers to the developing region with these transfers taking the form of adaptation measures. Transfers are conditional on the contribution of the developing region to mitigation,

---

[11]Conditional transfers in international climate policy are provided, for example, via the Global Environmental Facility which collects funds from developed countries and channels these towards developing countries.

[12]$p_i^{e,k}$ with $k = m, s$ and $i = I, D$ denotes the effective price in country i under matching m or subsidies s.

i.e., if the developing region provides x_D units of mitigation, the industrialized region matches this provision by adaptation financing of tx_D where $1 > t > 0$ is the so-called matching rate. In the subsequent parts of this chapter, we denote the policy scheme that matches mitigation with adaptation support shortly as "matching scheme",[13] while we denote the scheme that matches mitigation with monetary transfers (subsidies) as "subsidy scheme".[14] We assume throughout that $s = t$, i.e., transfer payments per unit of x_D are the same under subsidy and matching.

Adaptation is assumed to enhance the amount of effective private consumption as follows:

$$c_D = y_D + \delta t x_D. \quad (5)$$

The monetary input tx_D to the transfer scheme purchases adaptation a_D,[15] which in turn produces additional units of private characteristic consumption. $\delta > 0$ measures the productivity of adaptation transfers in generating additional private characteristic consumption. We assume that the unit price of adaptation a_D provided via international transfers — like the unit prices of all other goods — is equal to unity, such that we can write $a_D = tx_D$.

Furthermore, we assume that the relationship between adaptation and the effectively consumable amount X_D^c of mitigation in the

[13] In contrast to most applications or extensions of the standard matching model which was first developed by Guttman [1978], we do not consider mutual but one-sided matching. Unlike conventional matching models either considering the subsidization of public good purchases (e.g., Danziger and Schnytzer [1991] and Althammer and Buchholz [1993]) or the conditional contribution to the public good (e.g., Guttman [1987]), we consider the conditional contribution to another agent's private good "adaptation".

[14] Buchholz *et al.* [2011: F4] point to the equivalence of matching schemes based on conditional public good supply and specific schemes based on subsidy payments.

[15] As pointed out by, e.g., Ebert and Welsch [2012], adaptation might involve physical measures with different physical metrics. To make these metrics comparable, we adopt the common assumption that the monetary expenditure for aggregate adaptation reflects the efficient mix of adaptation measures (see also Ebert and Welsch [2012]).

developing region can be described by

$$X_D^c = X_D^c(X, a_D) = X - \gamma t x_D, \tag{6}$$

where $\gamma > 0$ is a constant parameter reflecting the mitigation-productivity diminishing effect of adaptation investments. The negative relationship between X_D^c and $a_D = tx_D$ reflects the assumed substitutive relationship between mitigation and adaptation that was described at the beginning of the model introduction. Hence, the input tx_D to the adaptation transfer scheme brings about a decline in the effectively consumable mitigation output by $\gamma t x_D$.

In the case of matching, maximization of welfare is again subject to the income constraint (2a), as in-kind adaptation transfers work through the change in effectively consumable outputs and thus through the arguments of the welfare function (1) only. Consequently, while subsidization in the previous subsection led to a modification of the developing region's budget constraint, financing adaptation affects the welfare function directly through its arguments X_D^c and c_D.

Nevertheless, the effects of adaptation can — in a similar way to the effects in the subsidy scenario — be interpreted as effective price changes. However, in contrast to the subsidy scenario, we have to distinguish between the effective prices of goods and of characteristics. In the subsidy case, amounts of goods and characteristics were supposed to coincide, i.e., $c_D = y_D$ and $X_D^c = X$, and consequently also the effective prices of goods and characteristics were the same.

However, in the matching case, amounts of goods and characteristics differ. While the effective price per unit of the public good x_D expressed in quantities of the private good y_D foregone remains unchanged, the effective price of consumable units of x_D^c is modified. As we now consider effective prices in the characteristics dimension, the effective price $p_D^{e,m}$ of purchasing a unit x_D^c has to reflect the amount of c_D that has to be foregone in exchange.

From the first-order conditions obtained from the maximization of (1) subject to (2a), (5) and (6), we derive

$$\frac{\frac{\partial U_D}{\partial X_D^c}}{\frac{\partial U_D}{\partial c_D}} = \frac{1 - t\delta}{1 - t\gamma}. \tag{7}$$

The RHS of (7) displays the effective price (in terms of c_D that has to be foregone in exchange) of purchasing one unit of x_D^c, i.e., $p_D^{e,m} = \frac{1-t\delta}{1-t\gamma}$ in the matching case. As we face multiple constraints we have to deal with a so-called "points rationing" problem [Cornes and Sandler, 1996: 259].[16]

Similar to the subsidy payment (see (4)), the in-kind-transfer reduces the numerator of the term on the RHS of (7). In contrast to the subsidy, however, the adaptation transfer also reduces the denominator of the RHS of (7). In summary, we observe two opposing effects of adaptation transfers: On the one hand, adaptation transfers "subsidize" public good provision by generating additional effective private consumption and thus make investments in public good provision more attractive. On the other hand, adaptation transfers reduce the productivity of public goods in generating domestic public benefits which in turn raises the attractiveness of private good purchases relative to public good contributions. It is clear that for $\gamma < \delta$ the effective price of the public good decreases due to the conditional adaptation transfer while it increases for $\gamma > \delta$. For $\gamma = \delta$, the two opposing effects on the effective price exactly compensate each other. In this case, matching would have no effect on the provision of the public good.

4. Comparison between Subsidies and Matching

In the following, the effects of subsidy and matching schemes on the welfare of the two regions are compared. For this comparison, we also have to consider the impact that the different policies have on the external benefits that arise from the provision of the global public good. To accomplish this, we consider the full income, E_i, of each region and not only its monetary income. A region's full income contains its monetary income plus the monetary value of the externalities received from the other region's public good provision

$$E_i = I_i + p_i^{e,k} x_j = I_i + p_i^{e,k}(X - x_i) = y_i + p_i^{e,k} X, \tag{8}$$

[16]To ensure for effective prices to be positive, we impose the parameter restrictions: $t\delta, t\gamma < 1$.

with $i, j = D, I$, $i \neq j$ and $k = m, s$ (for the use of the term "full income" in the public good context, also see Cornes and Sandler [1994: 409]; in the rationing context, Neary and Roberts [1980] and Rothbarth [1940–41] employ virtual magnitudes and Cornes and Sandler [1996: 262] use the term "virtual income").[17]

For the further analysis, it proves to be convenient to translate the welfare maximization problem in (1) and (2a) into an expenditure minimization problem with a region minimizing the cost of attaining welfare level $\bar{U}_i$ at given prices.

For the developing region, this gives

$$\min_{y_D, x_D} E_D = c_D + p_D^{e,k} X^C = I_D + p_D^{e,k} x_I \quad \text{subject to } U_D = \bar{U}_D.^{18} \tag{9}$$

From the first-order conditions of (9), the developing region's compensated demand $X_D (\equiv \partial E_D / \partial p_D^{e,k})$ for the public good

$$X_D(\bar{U}_D, p_D^{e,k}) \tag{10a}$$

can be derived (see Cornes [1992: 70] and Ihori [1996: 142]). Furthermore, we can express the optimal expenditure of region D as a function of the effective price and the region's welfare level, $E_D(\bar{U}_D, p_D^{e,k})$.

Analogously, we can proceed with respect to the industrialized region whose full income is given by (8) for $i = I$ and $p_I^{e,k} = 1$ throughout as there are no transfers channelled towards the industrialized world. Expenditure minimization then gives the compensated demand for X of the industrialized region

$$X_I(\bar{U}_I, p_I^{e,k}) = X_I(\bar{U}_I, 1) \tag{10b}$$

and its optimal expenditure $E_I(\bar{U}_I, 1)$ for a given welfare level and given prices.

[17] Put differently, the full income indicates the hypothetical income required for purchasing the amount of all consumed characteristics, i.e., including the amount of public characteristics provided by the other region.

[18] For the derivation of (9), consider that the developing region's full income consists of $E_D = c_D + p_D^{e,k} X^C = y_D + x_D t\delta + p_D^{e,k}[x_D(1 + t\gamma) + x_I] = I_D - (1 - \frac{1-\delta t}{1+\gamma t}(1 + t\gamma) - t\delta)x_D + \frac{1-\delta t}{1+\gamma t} x_I = I_D + \frac{1-\delta t}{1+\gamma t} x_I$.

In the global equilibrium, the aggregate expenditure of the two regions has to equal their aggregate full income and the demand for the global public good has to be equalized across regions. Multiplying the industrialized region's expenditure function by $p_D^{e,k}$, the developing region's expenditure function by $p_I^{e,k} = 1$ and summing up the resulting functions gives

$$E_D(\bar{U}_D, p_D^{e,k}) + p_D^{e,k} E_I(\bar{U}_I, 1) = p_D^{e,k} I_I + I_D + p_D^{e,k} X, \tag{11}$$

where

$$X = X_D(\bar{U}_D, p_D^{e,k}) = X_I(\bar{U}_I, 1) \tag{12}$$

holds with respect to global public good provision. From (11) and (12), we see that the interconnectedness of the two regions via the equilibrium condition for the global public good, (12), reduces the region's degree of freedom by one dimension.[19] Consider, for example, the effects of a policy-induced change in the effective price, $p_D^{e,k}$. If the two regions could independently decide upon their demand for the two goods, each region's expenditure minimization problem (i.e., (9) for the developing region) would independently give the optimal change in the respective consumption bundles for given welfare levels $\bar{U}_i$. In a world, in which each region is affected by the other region's preference for the global public good, however, a new equilibrium allocation can only be attained if the welfare levels in the two regions are allowed to adjust as well. It is these policy-induced welfare effects on which the focus of the following sections will be.

We assume that in the initial situation, no support of mitigation in the developing region is provided, i.e., in both regions the effective price of generating additional units of the public good is equal to unity.

[19]Please note that we focus on interior solutions in this chapter where the contribution of both regions to the public good is positive.

4.1. *Subsidy scheme*

We know from (4) that — in the subsidy scenario — the effective price of generating additional units of the public good in the developing region is

$$p_D^{e,s} = 1 - s \tag{13}$$

which implies

$$\frac{dp_D^{e,s}}{ds} = -1. \tag{14}$$

A marginal increase in the subsidy rate affects the industrialized region's monetary income as follows:

$$dI_I = -x_D ds. \tag{15}$$

To analyse the effects of subsidization on the two regions, we derive from (11):

$$\begin{aligned}\frac{\partial E_D}{\partial U_D} dU_D + p_D^{e,s} \frac{\partial E_I}{\partial U_I} dU_I &= -E_I \frac{dp_D^{e,s}}{ds} ds + I_I \frac{dp_D^{e,s}}{ds} ds + p_D^{e,s} dI_I \\ &\quad + p_D^{e,s} \frac{\partial X_I}{\partial U_I} dU_I \\ &= E_I ds - I_I ds + p_D^{e,s} dI_I + p_D^{e,s} \frac{\partial X_I}{\partial U_I} dU_I\end{aligned} \tag{16}$$

where we have replaced $\bar{U}_i$ by U_i to indicate that welfare levels adjust endogenously to subsidy induced price changes.

Furthermore, it follows from (12) that

$$\frac{\partial X_I}{\partial U_I} dU_I - \frac{\partial X_D}{\partial U_D} dU_D = -\frac{\partial X_D}{\partial p_D^{e,s}} ds. \tag{17}$$

Gathering (15), (16) and (17), we get

$$\begin{bmatrix} p_D^{e,s} \frac{\partial c_I}{\partial U_I} & \frac{\partial E_D}{\partial U_D} \\ \frac{\partial X_I}{\partial U_I} & -\frac{\partial X_D}{\partial U_D} \end{bmatrix} \begin{bmatrix} dU_I \\ dU_D \end{bmatrix} = \begin{bmatrix} (E_I - I_I) - p_D^{e,s} x_D \\ -\frac{\partial X_D}{\partial p_D^{e,s}} \end{bmatrix} ds \tag{18}$$

where $\frac{\partial E_i}{\partial U_i} \equiv p_i^{e,s}\frac{\partial X_i}{\partial U_i} + \frac{\partial c_i}{\partial U_i}$, $\frac{\partial X_i}{\partial U_i} > 0$ and where $c_i(U_i, p_i^{e,s})$ is the compensated demand function for private consumption in region $i(i = D, I)$ with $\frac{\partial c_i}{\partial U_i} > 0$.

From (18), we can now derive the welfare effects of introducing a mitigation subsidy. For the industrialized region, we obtain:

$$\frac{dU_I}{ds} = \frac{\left[(I_I - E_I) + p_D^{e,s}x_D\right]\frac{\partial X_D}{\partial U_D} + \frac{\partial X_D}{\partial p_D^{e,s}}\frac{\partial E_D}{\partial U_D}}{\Delta_s}$$
$$= \frac{\left[x_D(p_D^{e,s} - 1)\right]\frac{\partial X_D}{\partial U_D} + \frac{\partial X_D}{\partial p_D^{e,s}}\frac{\partial E_D}{\partial U_D}}{\Delta_s}$$
$$= \frac{p_D^{e,s}x_D\frac{\partial X_D}{\partial U_D}}{\Delta_s} - \frac{x_D\frac{\partial X_D}{\partial U_D} - \frac{\partial E_D}{\partial U_D}\frac{\partial X_D}{\partial p_D^{e,s}}}{\Delta_s}, \tag{19}$$

where $\Delta_s = -p_D^{e,s}\frac{\partial c_I}{\partial U_I}\frac{\partial X_D}{\partial U_D} - \frac{\partial E_D}{\partial U_D}\frac{\partial X_I}{\partial U_I} < 0$. In the initial situation where no support of mitigation in the developing region is provided, it holds that $\Delta_s = -(\frac{\partial c_I}{\partial U_I}\frac{\partial X_D}{\partial U_D} + \frac{\partial E_D}{\partial U_D}\frac{\partial X_I}{\partial U_I})$ which we additionally label for later use Δ^0. The term $\frac{p_D^{e,s}x_D\frac{\partial X_D}{\partial U_D}}{\Delta_s}$ in (19) is the effect of the decline in income I_I which is due to the payment of the subsidy. $-\frac{x_D\frac{\partial X_D}{\partial U_D} - \frac{\partial E_D}{\partial U_D}\frac{\partial X_D}{\partial p_D^{e,s}}}{\Delta_s}$ reflects the spillover effect from a decrease in the effective price $p_D^{e,s}$ on the industrialized region. It includes an income and a substitution price effect.

The total income effect arising from the introduction of the subsidy is thus given by:

$$TIE_{sub}^I = [p_D^{e,s} - 1]\frac{x_D\frac{\partial X_D}{\partial U_D}}{\Delta_s}. \tag{20}$$

The sign of the total income effect depends on the level of the pre-policy effective price of mitigation, $p_D^{e,s}$, and the change in the effective price induced by the policy — which in the case of a subsidy is equal to one ($\frac{\partial p_D^{es}}{\partial s} = 1$). For the intuition behind this result, consider the composition of the total income effect. On the one hand, as stated above, the subsidy payment from the industrialized region to the developing region causes a direct negative income effect to region I. On the other hand, the induced increase in mitigation in the developing region results in a positive spillover effect on the full income of region I. The strength of the latter effect depends on the policy-induced change in the effective price. The stronger the change in the effective price, the higher the increase in the mitigation of the developing region and the higher the spillover income effect. For the spillover income effect to exactly offset the negative, transfer-induced income effect in region I when the effective pre-policy prices p_i^e are equal, the conditional transfer has to lead to a one-to-one decrease in region D's effective price. In this case, the total income effect is equal to zero. For a policy-induced price reduction which is higher (lower), the total income effect would become positive (negative).

From (20), we see that subsidization leads to a one-to-one reduction of p_D^e. Assuming that no policy is conducted at the outset, the effective prices in the two regions are equal to unity initially. Thus, the total income effect is equal to zero.[20]

From (19), the substitution price effect that spills over to the industrialized region is given by:

$$SPE_{sub}^{I} = \frac{\frac{\partial X_D}{\partial p_D^{e,s}} \frac{\partial E_D}{\partial U_D}}{\Delta_s} > 0. \tag{21}$$

Due to the subsidy-induced reduction in its effective price of mitigation, the developing region increases its contribution to mitigation which has, due to the global nature of the public good mitigation, a positive effect on welfare of region I. So, overall the subsidy payment and associated decrease in region D's effective price raise the industrialized region's welfare (also see e.g., Bergstrom [1989] and Ihori

[20]Please note that this result does not depend on the absolute level of the pre-policy prices, $p_i^{e,s}$, but on the assumption that the pre-policy prices and thus the productivities of mitigation across regions are identical. For $p_i^{e,s} \neq 1$ the term in brackets in (20) would read $[p_D^{e,s} - \frac{\partial p_D^{es}}{\partial s} p_I^{e,s}]$. Thus, if prices across regions differ ($p_D^{e,s} \neq p_I^{e,s}$), our result modifies accordingly. If, for example, $p_D^{e,s} > 1$ ($p_D^{e,s} < 1$), the decrease in the effective price that renders the total income effect zero, has to be higher (lower). We consider a case in which pre-policy prices differ in Section 4.3.

[1996] for such positive welfare effects on the payer of conditional transfers).

For the developing region, the welfare effect of the subsidy introduction is given by (see (18)):

$$\frac{dU_D}{ds} = \frac{p_D^{e,s} x_D \frac{\partial X_I}{\partial U_I}}{\Delta_s} - \frac{x_D \frac{\partial X_I}{\partial U_I} + p_D^{e,s} \frac{\partial c_I}{\partial U_I} \frac{\partial X_D}{\partial p_D^{e,s}}}{\Delta_s}. \tag{22}$$

$\frac{p_D^{e,s} x_D \frac{\partial X_I}{\partial U_I}}{\Delta_s}$ shows the effect which the transfer-induced decline of monetary income in the industrialized region has on welfare in the developing region. Predictably, this effect is negative as the reduction in income reduces the industrialized region's contribution to mitigation. The term $-\frac{x_D \frac{\partial X_I}{\partial U_I} + p_D^{e,s} \frac{\partial c_I}{\partial U_I} \frac{\partial X_D}{\partial p_D^{e,s}}}{\Delta_s}$ reflects the effect that results from the decrease in the effective price of public good provision $p_D^{e,s}$ for the developing region. Again this effect encompasses an income and a substitution price effect.

The total income effect for the developing region reads

$$TIE_{sub}^D = [p_D^{e,s} - 1] \frac{x_D \frac{\partial X_I}{\partial U_I}}{\Delta_s}, \tag{23}$$

where, as in the case of the industrialized region, the sign of the total income effect depends on the policy-induced change in the effective price. In the case of subsidization, we again get the result of two income effects that perfectly offset each other.

The substitution price effect for the developing region

$$SPE_{sub}^D = -\frac{p_D^{e,s} \frac{\partial c_I}{\partial U_I} \frac{\partial X_D}{\partial p_D^{e,s}}}{\Delta_s} < 0 \tag{24}$$

is, however, negative as the region's substitution-induced increase in public good provision is met by a decrease in the provision of the other region. Thus, the overall welfare effect of the subsidy for the transfer-receiving region is negative.

Summing up, we show that the transfer-paying industrialized region gains welfare while the receiving developing region loses welfare. The conditional monetary transfer raises the developing region's demand for and supply of the public good. The rise in the developing region's public-good supply — in turn — is like a transfer in-kind to the industrialized world and, as we assumed that both private and public characteristic consumption behave like normal goods, "increasing such a transfer induces a higher demand for private consumption and thus a lower contribution to the public good" (for this explanation and quote see Buchholz and Peters, 2005: 29) in the industrialized region.

4.2. *Matching scheme: Reference scenario with* $\delta = 1$

When considering the effects of matching transfers on welfare, it proves to be helpful to first devote special attention to the case in which the productivity of adaptation in generating additional units of the consumable private good c_D is equal to unity, i.e., $\delta = 1$. In this case, the monetary input to the adaptation transfer scheme is equal to its output ($tx_D\delta = tx_D = a_D$). In the course of Section 4.2, we largely focus on this simplified case while the case of $\delta > 1$ is dealt with in more detail in Section 4.3.

4.2.1. *Effects of matching transfers on the industrialized region's welfare*

To analyze the influence of matching transfers on the welfare of the transfer paying industrialized region, we proceed in a stepwise fashion: In step one, we regard a simplified case in which adaptation has no "productivity effect" on mitigation, i.e., $\gamma = 0$ and therefore $X_D^c = X$ ("*Case 1*"). Step two then comprises the analysis of the complete, i.e., the "real" case where all adaptation effects are taken into account, i.e., $\gamma \neq 0$ and $X_D^c = X - \gamma tx_D$ ("*Case 2*"). This stepwise procedure allows us to get a better intuition on the differential effects of matching transfers versus subsidy payments.

Case 1: In this first matching case, the developing world faces the effective price $p_D^{e,m'} = 1 - t\delta$ of public characteristic generation. The corresponding expenditure function is given by

$$E_D = I_D + (1 - t\delta)x_I. \tag{25a}$$

It holds that

$$\frac{\partial E_D}{\partial U_D}dU_D + p_D^{e,m'}\frac{\partial E_I}{\partial U_I}dU_I = E_I\delta dt - I_I\delta dt + p_D^{e,m'}dI_I + p_D^{e,m'}\frac{\partial X_I}{\partial U_I}dU_I. \tag{26a}$$

A marginal change in the transfer rate affects the industrialized region's monetary income in the same way as the change in the subsidy rate in Section 4.1 (see (15)):

$$dI_I = -x_D dt. \tag{27}$$

In equivalence to (18), welfare changes due to the introduction of a matching scheme can now be derived from (17), (26a) and (27):

$$\begin{bmatrix} p_D^{e,m'}\frac{\partial c_I}{\partial U_I} & \frac{\partial E_D}{\partial U_D} \\ \frac{\partial X_I}{\partial U_I} & -\frac{\partial X_D}{\partial U_D} \end{bmatrix}\begin{bmatrix} dU_I \\ dU_D \end{bmatrix} = \begin{bmatrix} \delta(E_I - I_I) - p_D^{e,m'}x_D \\ -\delta\frac{\partial X_D}{\partial p_D^{e,m'}} \end{bmatrix} dt. \tag{28a}$$

We get for the industrialized region:

$$\frac{dU_I}{dt} = \frac{p_D^{e,m'}x_D\frac{\partial X_D}{\partial U_D}}{\Delta_1} - \frac{\delta\left[x_D\frac{\partial X_D}{\partial U_D} - \frac{\partial E_D}{\partial U_D}\frac{\partial X_D}{\partial p_D^{e,m'}}\right]}{\Delta_1} > 0, \tag{29a}$$

where $\Delta_1 = -p_D^{e,m'}\frac{\partial c_I}{\partial U_I}\frac{\partial X_D}{\partial U_D} - \frac{\partial E_D}{\partial U_D}\frac{\partial X_I}{\partial U_I} < 0$. In the initial situation where no support of mitigation in the developing region is provided, $\Delta_1 = \Delta^0$ holds.

The term $\frac{p_D^{e,m'}x_D\frac{\partial X_D}{\partial U_D}}{\Delta_1}$ in (29a) reflects — comparable to (19) in the subsidy case — the effect of a decline in income I_I, now due to the expenditure for the in-kind transfer. Also as in (19), $-\frac{\delta[x_D\frac{\partial X_D}{\partial U_D} - \frac{\partial E_D}{\partial U_D}\frac{\partial X_D}{\partial p_D^{e,m'}}]}{\Delta_1}$ reflects the spillover effect of the decrease in the developing region's effective price of public good provision, now $p_D^{e,m'}$, which again results in an income and a substitution price effect.

The total income effect and the substitution price effect in the industrialized region are now given by

$$TIE^{I,C1}_{match} = [p^{e,m'}_D - \delta]\frac{x_D\frac{\partial X_D}{\partial U_D}}{\Delta_1}, \tag{30a}$$

$$SPE^{I,C1}_{match} = \delta\frac{\frac{\partial X_D}{\partial p^{e,m'}_D}\frac{\partial E_D}{\partial U_D}}{\Delta_1} > 0. \tag{31a}$$

If one unit of adaptation induces one additional unit of private consumption, i.e., $\delta = 1$, we see from comparing (20) and (30a) as well as (21) and (31a) respectively that in this case the subsidy scheme and the matching scheme are equivalent, such that the matching transfer is unambiguously and equally welfare enhancing for the industrialized region as the subsidy.

Case 2: For $\gamma \neq 0$, the effective price of providing an additional unit of the public characteristic, $p^{e,m}_D = \frac{1-t\delta}{1-t\gamma}$, is higher than the effective price in Case 1, i.e., $p^{e,m}_D > p^{e,m'}_D$. The adverse mitigation productivity effect of adaptation implies a lower characteristic level X^c_D obtained from the public good X in the developing region. This lower productivity translates into a higher effective price for raising the consumption level of the public characteristic in the developing region.[21] We now face a case similar to the one analyzed by Ihori [1996: 146–149] where different productivities in producing a public good prevail in subsidizing and subsidized regions.

The modified expenditure function of the developing region now reads

$$E_D = I_D + \frac{1-t\delta}{1-t\gamma}x_I. \tag{25b}$$

[21]Conversely, private ancillary benefits of climate protection are frequently modelled to offset the cost of respective protection activities and thus to reduce the effective price of climate protection (see e.g., Ekins [1996: 163] and Rübbelke [2002: 105]).

(26a) changes accordingly to

$$\frac{\partial E_D}{\partial U_D}dU_D + p_D^{e,m}\frac{\partial E_I}{\partial U_I}dU_I = E_I\frac{\delta-\gamma}{(1-t\gamma)^2}dt - I_I\frac{\delta-\gamma}{(1-t\gamma)^2}dt + p_D^{e,m}dI_I + p_D^{e,m}\frac{\partial X_I}{\partial U_I}dU_I \tag{26b}$$

and gathering (17), (26b) and (27) gives

$$\begin{bmatrix} p_D^{e,m}\frac{\partial c_I}{\partial U_I} & \frac{\partial E_D}{\partial U_D} \\ \frac{\partial X_I}{\partial U_I} & -\frac{\partial X_D}{\partial U_D} \end{bmatrix} \begin{bmatrix} dU_I \\ dU_D \end{bmatrix} = \begin{bmatrix} (E_I - I_I)\frac{\delta-\gamma}{(1-t\gamma)^2} - p_D^{e,m}x_D \\ -\frac{\partial X_D}{\partial p_D^{e,m}}\frac{\delta-\gamma}{(1-t\gamma)^2} \end{bmatrix} dt. \tag{28b}$$

From (28b), we obtain

$$\frac{dU_I}{dt} = \frac{p_D^{e,m}x_D\frac{\partial X_D}{\partial U_D}}{\Delta_2} - \frac{\frac{\delta-\gamma}{(1-t\gamma)^2}\left[x_D\frac{\partial X_D}{\partial U_D} - \frac{\partial E_D}{\partial U_D}\frac{\partial X_D}{\partial p_D^{e,m}}\right]}{\Delta_2}, \tag{29b}$$

where $\Delta_2 = -p_D^{e,m}\frac{\partial c_I}{\partial U_I}\frac{\partial X_D}{\partial U_D} - \frac{\partial E_D}{\partial U_D}\frac{\partial X_I}{\partial U_I} < 0$. Initially, when no support of mitigation in the developing region is provided, we again have $\Delta_2 = \Delta^0$.

The interpretation of the different components of (29b) follows the same lines as the interpretation of (29a). Equivalently, the modified expressions for the total income and the substitution price effect now read

$$TIE_{match}^{I,C2} = \left[p_D^{e,m} - \frac{\delta-\gamma}{(1-t\gamma)^2}\right]\frac{x_D\frac{\partial X_D}{\partial U_D}}{\Delta_2}, \tag{30b}$$

$$SPE_{match}^{I,C2} = \frac{\delta-\gamma}{(1-t\gamma)^2}\frac{\frac{\partial X_D}{\partial p_D^{e,m}}\frac{\partial E_D}{\partial U_D}}{\Delta_2} > 0, \tag{31b}$$

where the sign of the total income effect and the sign of substitution price effect crucially depend on $\gamma \gtrless \delta$. For $\gamma > \delta$, the total income effect and the substitution price effect are negative, leaving the industrial region unambiguously worse off. For $\gamma < \delta$, however, the income effect is ambiguous while the substitution price effect is positive.

Discussion of Cases 1 and 2: As we can easily observe from a comparison between (29a) and (29b), no incentive arises for the industrialized region to ever choose a matching transfer instead of a direct mitigation subsidy if $\gamma > \delta$.

For $\gamma < \delta$, the implications of the negative mitigation-productivity effect of adaptation for the industrialized region's welfare are ambiguous as $\delta \lesseqgtr \frac{\delta-\gamma}{(1-t\gamma)^2}$, (recall that $\Delta_1 = \Delta_2 = \Delta^0$ initially). Compared to Case 1, the adverse mitigation-productivity effect reduces welfare in the industrialized region under this parameter constellation if

$$\gamma > \frac{2t\delta - 1}{t^2\delta}, \quad \text{resp. } \gamma > \frac{2t-1}{t^2} \quad \text{for } \delta = 1. \tag{32}$$

If (32) holds, the spillover income and substitution price effects are lower than in the subsidization scenario and not only the total income effect but also the total welfare effect can become negative (see (29b)). For this to happen, the positive substitution price effect has to be dominated by the now negative total income effect which is more likely if $\frac{\partial E_D}{\partial U_D}\frac{\partial X_D}{\partial p_D^{e,m}}$ is small while $x_D\frac{\partial X_D}{\partial U_D}$ is large.[22] If, however, (32) does not hold, the industrialized region's welfare gain is even higher than in Case 1. For $\delta = 1$, when Case 1 and the subsidy case are equivalent, the subsidy scheme is preferred by the industrialized region if (32) holds while matching is preferred if (32) does not hold.

Inequality (32) tends to hold with large mitigation productivity losses induced by adaptation (which is reflected by a large γ). This is in accordance with intuition: In the case of a high mitigation productivity decline, the policy-induced change in mitigation productivity is relatively low such that the spillover income effect from the increase in region D's mitigation is not sufficient to compensate the negative income effect from the matching transfer.

Also in accordance with intuition, (32) tends to hold for low values of t and δ, as this implies a low effective-price reducing influence of the transfer (see RHS of (7) and take into account that we

[22]The total welfare effect becomes negative, if $(1+(t\gamma-2)t)\gamma x_D\frac{\partial X_D}{\partial U_D} > (\gamma-1)\frac{\partial E_D}{\partial U_D}\frac{\partial X_D}{\partial p_D^{e,m}}$. A necessary condition for this to happen is that $1+(t\gamma-2)t > 0$ which is equivalent to condition (32).

still assume $\delta > \gamma$).[23] In this case, the price-reducing influence of a low transfer rate t cannot countervail the adverse price effect of the mitigation productivity loss. If t is, however, sufficiently high, the reduction in the effective price that results from increasing the transfer rate might even be higher in the presence of feedback effects on mitigation than without.

Summarizing the results for $\delta = 1$ and $\delta > \gamma$, the industrialized region always gains in Case 1 — like in the equivalent subsidy setting — from its payment of transfers towards the developing region. The influence of the mitigation-productivity diminishing effect of adaptation on the welfare change induced by transfers is, however, ambiguous (Case 2). Surprisingly, the adaptation-induced adverse effect on the productivity of mitigation in the developing region might even be beneficial from the transfer payer's point of view. It is more likely that this happens, when t and δ are high and the adverse effect of adaptation on mitigation productivity (reflected by a small γ) is low. Then, the industrialized region will prefer the use of the matching scheme over subsidization.

4.2.2. *Effects of matching transfers on the developing region's welfare*

Again, we assume in a first step that adaptation has no productivity effect on mitigation, i.e., $X_D^c = X$. Thereafter, in a second case, we consider the situation where $X_D^c = X - tx_D\gamma$.

Case 1: As in the discussion of Case 1 from the industrialized region's point of view, the effective price of the public good for the developing region is $p_D^{e,m'} = 1 - t\delta$.

From (28a), we now derive the effect of the matching transfer on the developing region's welfare

$$\frac{dU_D}{dt} = \frac{p_D^{e,m'} x_D \frac{\partial X_I}{\partial U_I}}{\Delta_1} - \frac{\delta \left[x_D \frac{\partial X_I}{\partial U_I} + p_D^{e,m'} \frac{\partial c_I}{\partial U_I} \frac{\partial X_D}{\partial p_D^{e,m'}} \right]}{\Delta_1} \tag{33a}$$

[23]Differentiation of the RHS of (32) with respect to δ resp. t yields positive values (recall that $1 - \delta t > 0$).

from which the total income effect and the substitution price effect are given by

$$TIE^{D,C1}_{match} = [p^{e,m'}_D - \delta]\frac{x_D \frac{\partial X_I}{\partial U_I}}{\Delta_1}, \tag{34a}$$

$$SPE^{D,C1}_{match} = -\delta \frac{p^{e,m'}_D \frac{\partial c_I}{\partial U_I} \frac{\partial X_D}{\partial p^{e,m'}_D}}{\Delta_1} < 0. \tag{35a}$$

For $\delta = 1$, the matching scheme and the subsidy scheme are once more equivalent which implies that the matching transfer unambiguously reduces welfare in the developing region as, again, the total income effect is zero while the substitution price effect is negative.

Case 2: The effective price in this case is again $p^{e,m}_D = \frac{1-t\delta}{1-t\gamma}$. From (28b), the overall effect of a change in the transfer rate is now equal to

$$\frac{dU_D}{dt} = \frac{p^{e,m}_D x_D \frac{\partial X_I}{\partial U_I}}{\Delta_2} - \frac{\frac{\delta-\gamma}{(1-t\gamma)^2}\left[x_D \frac{\partial X_I}{\partial U_I} + p^{e,m}_D \frac{\partial c_I}{\partial U_I} \frac{\partial X_D}{\partial p^{e,m}_D}\right]}{\Delta_2} \tag{33b}$$

with the total income effect and the substitution price effect equalling:

$$TIE^{D}_{match} = \left[p^{e,m}_D - \frac{\delta-\gamma}{(1-t\gamma)^2}\right]\frac{x_D \frac{\partial X_I}{\partial U_I}}{\Delta_2}, \tag{34b}$$

$$SPE^{D}_{match} = -\frac{\delta-\gamma}{(1-t\gamma)^2}\frac{p^{e,m}_D \frac{\partial c_I}{\partial U_I} \frac{\partial X_D}{\partial p^{e,m}_D}}{\Delta_2} < 0. \tag{35b}$$

The interpretation of (33b)–(35b) is equivalent to the interpretation of (29b)–(31b), i.e., Case 2 for the industrialized region. As in (30b), the sign of the total income effect is ambiguous for $\gamma < \delta$ while the substitution price effect for the developing region is again negative. Whether or not the overall welfare effect is negative depends, as for the industrialized region, on the strength of the transfer-induced effects on consumption and mitigation as well as on the transfer rate. For $\gamma > \delta$, the total income effect becomes negative — as was the case for the industrialized region — but the substitution price effect now becomes positive, resulting in an ambiguous welfare effect. Whether or not the overall welfare effect is positive or negative depends crucially on the relative strength of income and substitution price effects.

Discussion of Cases 1 and 2: Regarding Cases 1 and 2, we can easily see from a comparison of (34a) and (34b) as well as of (35a) and (35b) respectively, that while matching always induces a negative welfare effect for the developing region in Case 1, it might have a positive welfare effect in Case 2.

If $\gamma > \delta$, whether or not this would happen, depends, as already said, on the relative strength of income and substitution price effects. If the substitution price effect is stronger than the income effect that is induced by the price change (i.e., if $p_D^{e,k} \left| \frac{\partial c_I}{\partial U_I} \frac{\partial X_D}{\partial p_D^{e,k}} \right| > x_D \frac{\partial X_I}{\partial U_I}$), the welfare in Case 2 is higher than in Case 1. If the substitution price effect is sufficiently strong, the overall welfare effect could even become positive — which can never arise under subsidization.[24]

If $\gamma < \delta$, the change of the developing region's welfare due to the feedback effect of adaptation on mitigation depends — as in the case of the industrialized region — on $\delta \gtrless \frac{\delta - \gamma}{(1-t\gamma)^2}$. In contrast to the case of the industrialized region, however, the relative strengths of income and substitution price effects also matter for the developing region. According to (33a) and (33b), welfare is reduced if

$$\gamma > \frac{2t\delta - 1}{t^2\delta}, \quad \text{and} \quad x_D \frac{\partial X_I}{\partial U_I} > p_D^{e,k} \left| \frac{\partial c_I}{\partial U_I} \frac{\partial X_D}{\partial p_D^{e,k}} \right|, \quad \text{or} \tag{36a}$$

$$\gamma < \frac{2t\delta - 1}{t^2\delta}, \quad \text{and} \quad x_D \frac{\partial X_I}{\partial U_I} < p_D^{e,k} \left| \frac{\partial c_I}{\partial U_I} \frac{\partial X_D}{\partial p_D^{e,k}} \right|, \tag{36b}$$

with $p_D^{e,m'} = p_D^{e,m} (= 1)$ in the ex-ante situation and consequently $k = m', m$. If (36a) or (36b) holds for $\delta = 1$, the developing region would be better off under a subsidy scheme than under a matching scheme.

[24] As, however, the industrialized region would never choose matching over subsidization if $y > \delta$ as we have shown before, we abstain from the discussion of this case in the rest of the chapter.

How the welfare effect changes due to $\gamma \neq 0$ depends crucially on the relative sizes of substitution price effect and positive income effect — or more specifically, on the size of $p_D^{e,k}|\frac{\partial c_I}{\partial U_I}\frac{\partial X_D}{\partial p_D^{e,k}}|$ relative to $x_D\frac{\partial X_I}{\partial U_I}$. This is due to the fact that a change in the effective price shifts the positive income effect and the negative substitution effect in opposite directions. Take (36a), for example, in which the reduction in the effective price due to matching is lower than for $\gamma = 0$. Consequently, the positive income effect is smaller as the income-induced increase in global public good provision in the developing region is less strong. The negative substitution price effect, however, also decreases in strength. The lower price change induces a lower substitution driven increase in mitigation by region D and thus a lower matching decrease in mitigation of region I. As the latter hurts region D, a decrease in its strength is good for the developing region. Now, if the reduction in the positive income effect is larger than the decrease in the negative substitution price effect, region D's welfare falls even more in the matching than in the subsidy scenario.

Essentially, the comparative statics with respect to the model parameters in (36a) are the same for the developing region as in (32) for the industrialized region. The higher γ, the more likely welfare is to decrease (see (33a) and (33b)) while a positive effect on welfare becomes more likely if the adverse effect of adaptation on mitigation is small (reflected by a small γ). A high transfer rate t and a strong impact of adaptation on the benefits from private good consumption as reflected by a large δ (we discuss the latter case in more detail in Section 4.3) would also be beneficial when (36a) holds.

(36b) applies when the substitution price effect is stronger than the positive income effect, i.e., if $p_D^{e,k}|\frac{\partial c_I}{\partial U_I}\frac{\partial X_D}{\partial p_D^{e,k}}|$ is larger than $x_D\frac{\partial X_I}{\partial U_I}$. If this condition holds, a negative effect of $\gamma \neq 0$ on welfare becomes more likely if γ is low. In this case, the decrease in the effective price is strong enough to increase the positive income effect compared to the subsidy case, but at the same time, the negative substitution effect also becomes stronger. So, if the latter effect is dominating the former, welfare falls compared to the subsidy scenario. From a comparative statics perspective, the total adverse welfare effect now not only becomes stronger with lower levels of γ but also with higher levels of t and δ. In turn, a low transfer rate, a small positive effect of abatement on private consumption and a strong negative effect of abatement on mitigation would make a welfare reduction due to the adverse effect of adaptation on mitigation less likely.

To summarize, we surprisingly find that the decline in the productivity of mitigation in the developing region may have a positive welfare effect on this region. This resembles the results of Buchholz and Konrad [1994] who find in a multi-stage game that countries may have an incentive to choose an abatement technology which is more costly than others in the first stage of the game. The effect of matching on the welfare of both regions depends crucially on $\delta \gtrless \frac{\delta-\gamma}{(1-t\gamma)^2}$ and, in case of the developing region, on the relative size of substitution price and positive income effect. Depending on the parameterization of t, γ and δ, either region may win or lose. A matching scheme may become welfare enhancing for both regions if $\delta < \frac{\delta-\gamma}{(1-t\gamma)^2}$ and the positive income effect dominates the substitution price effect in region D.

4.3. *Comparison of policy schemes: Alternative technological productivities*

In the previous section, we largely concentrated on the special case in which an adaptation transfer unit enhanced the effective consumption of the private good by exactly the same amount as the purchase of another unit of the private good would have done. In this section, we relax this very specific assumption and take a look at more general constellations in which we allow not only δ but also the effective ex-ante price p_D^{e*}, i.e., the effective price before the payment of international transfers,[25] to differ from unity.

[25] Akai [2003] discusses the different implications of cost-differentials in public good provision generated by public policies on the one hand and cost differentials that already existed before policy implementation for income transfer policy on the other hand.

In the first scenario, we assume that the productivity δ of in-kind adaptation transfers in generating additional private consumption is higher than in the Reference Scenario discussed in Section 4.2, i.e., $\delta > 1$.

In a second scenario, we suppose that the effective price of the public good is ex ante, i.e., before a transfer is introduced, smaller than unity in the developing world.

Throughout the analysis of both scenarios, we keep the assumption that $\delta > \gamma$, i.e., adaptation transfers cause effective costs of public good contributions to decline, and $t = s$, i.e., matching and subsidy rates are equal.

Alternative Scenario 1 – $\boldsymbol{\delta > 1}$, $\boldsymbol{p_D^{e*} = 1}$: In this scenario $t = s < t\delta$ and consequently the productivity of adaptation transfers in raising private good benefits is higher than in the Reference Scenario where δ is equal to unity. As a result, the reduction of the effective price due to matching transfers is stronger for $\delta > 1$ than for $\delta = 1$.

In Alternative Scenario 1, the total income effect in the industrialized region becomes more positive/less negative if δ rises. This result is independent of whether or not an effect of adaptation on mitigation arises, as we can see from (30a) and (30b). Yet, while the income effect becomes unambiguously positive in (30a), the sign of the total income effect in (30b) remains ambiguous. It only becomes positive (and hence more beneficial than in the subsidization setting), when the change in the effective price is higher than in the subsidy case, i.e., if $\frac{\delta-\gamma}{(1-t\gamma)^2} > 1$. Regarding the positive substitution price effect, it becomes stronger for increases in δ in Case 1 as well as in Case 2 (see (31a) and (31b)). Thus overall, a higher productivity of adaptation raises the utility that the industrialized region obtains from its adaptation transfers unambiguously and makes it more likely for matching contributions to become advantageous over subsidy payments. The latter holds throughout in Case 1 of the Alternative Scenario 1 (i.e., with $\delta > 1$ and $\gamma = 0$) as the decrease in the effective price is in this case always higher under matching than under subsidization. This finding is in line with Cornes and Hartley [2007: 215] who show that a reduction in an agent's unit cost of public good provision raises all other agents' welfare. This result also resembles the finding by Ebert and Welsch [2012] in a different two-country setting, that — conversely — improvements in the benefits from greenhouse gas emissions, and consequently an increase in the marginal cost of mitigation,[26] have negative effects on the other country's welfare. In our model, the benefits of mitigation activities x_D in the developing region are raised (and with it the effective price of mitigation characteristics reduced) through the receipt of adaptation transfers and thus, conversely, welfare in the industrialized region improves.

In the developing region — as in the industrialized region before — the total income effect rises with higher values of δ in Cases 1 and 2. Again, the total income effect is always positive for $\delta > 1$ in Case 1, while it only becomes positive for $\frac{\delta-\gamma}{(1-t\gamma)^2} > 1$ in Case 2. Also as in the industrialized region, the substitution price effect is stronger due to the stronger decrease in the effective price. Since the substitution price effect is negative for the developing region, this implies, however, a stronger drag on welfare. Whether a higher δ improves or worsens welfare in the developing world, depends, as in the Reference Scenario, on the relative strength of substitution price and income effects. For the developing region's welfare in the matching setting to exceed its welfare in the subsidy setting, the following conditions have to hold (see (22) and (33b)):

$$\frac{\delta-\gamma}{(1-t\gamma)^2} > 1, \quad \text{and} \quad x_D \frac{\partial X_I}{\partial U_I} > p_D^{e,k} \left| \frac{\partial c_I}{\partial U_I} \frac{\partial X_D}{\partial p_D^{e,k}} \right|, \quad \text{or} \tag{37a}$$

$$\frac{\delta-\gamma}{(1-t\gamma)^2} < 1, \quad \text{and} \quad x_D \frac{\partial X_I}{\partial U_I} < p_D^{e,k} \left| \frac{\partial c_I}{\partial U_I} \frac{\partial X_D}{\partial p_D^{e,k}} \right|, \tag{37b}$$

with $p_D^{e,m} = p_D^{e,s} (= 1)$ in the ex-ante situation and thus $k = m, s$. The economic intuition behind this result is essentially the same as for

[26]Ebert and Welsch [2012: 59] note that "The emission enhancing effect of productivity improvements is intuitive, since such improvements imply an increase in the benefits foregone when emissions are reduced, that is, an increase in marginal mitigation costs."

(36a) and (36b). Please note that in Case 1, $\frac{\delta-\gamma}{(1-t\gamma)^2} = \delta > 1$ always holds.

(37a) tells us that an increase in δ makes a positive welfare effect in the developing region more likely if $x_D \frac{\partial X_I}{\partial U_I} > p_D^{e,k} | \frac{\partial c_I}{\partial U_I} \frac{\partial X_D}{\partial p_D^{e,k}} |$, i.e., if the positive income effect exceeds the substitution price effect. Thus, efforts of the industrialized region to increase the effectiveness of adaptation support in this case also affect the developing region's welfare positively. However, if $x_D \frac{\partial X_I}{\partial U_I} < p_D^{e,k} | \frac{\partial c_I}{\partial U_I} \frac{\partial X_D}{\partial p_D^{e,k}} |$, the total welfare effect of matching is always negative and even stronger than under subsidization.

If (37b) holds, matching is only advantageous compared to subsidization from the developing region's point of view, if the transfer-induced decline in the effective public good price is lower than in the subsidy scheme, i.e., if $\frac{\delta-\gamma}{(1-t\gamma)^2} < 1$. As in this case a lower δ would be advantageous from the developing region's perspective, region D might find it paradoxically in its best interest to prevent δ from being too high that is to prevent adaptation support from being too efficient.

Summarizing, an improvement of the productivity (in raising private consumption) of adaptation support channelled towards the developing countries tends to raise the welfare in industrialized countries. Therefore, donor regions have a strong interest to disburse adaptation transfers in the way that most effectively raises the developing world's private consumption c_D. In contrast, developing regions may prefer a less effective adaptation "technology" generating a lower level of additional private characteristic consumption. Thus, we may face a principal-agent situation where the donor of transfers is the principal and the transfer receiving region is the agent and where there might be a risk that the agents have an incentive to waste or misguide adaptation resources.[27]

---

[27]It should be noted that the productivity of adaptation support in generating additional private consumption might also affect the impact of adaptation on mitigation benefits, i.e., $\frac{\partial X_D^c(X, a_D)}{\partial a_D}$. This might in turn strengthen or weaken potential incentives to waste adaptation resources.

Alternative Scenario 2 – $\delta \geq 1$, $p_D^{e*} < 1$: Regularly, cheaper climate change mitigation options are supposed to exist in developing countries as the present application of very polluting technologies in these countries can be overcome at lower cost than improving the already rather clean technologies in industrialized countries. In that sense, mitigation projects in developing countries tend to be more productive than those in industrialized countries.

In our model, a productivity advantage of the developing region in global public good provision can be translated into an effective ex-ante price p_D^{e*} smaller than 1. For the industrialized region, this implies higher total income effects from subsidies as well as matching transfers and thus higher policy-induced welfare gains, respectively lower policy-induced welfare losses (see (19), (29a) and (29b)). Thus, the transfer paying industrialized region unambiguously gains from supporting cheaper climate change mitigation in the developing region.

In the developing region, the higher mitigation productivity also has a positive influence on the total income effect under subsidization as well as matching (see (23), (34a) and (34b)). Simultaneously, the negative substitution price effect (see (24), (35a) and (35b)) is weakened by the lower ex-ante price p_D^{e*}. Despite the positive effects of higher mitigation productivity on income and substitution price effects in the transfer receiving region, it may, however, still lose welfare due to the industrialized region's subsidy or matching transfer payments. For the subsidy scheme, these results for the Alternative Scenario 2 are consistent with the earlier findings in Ihori [1996: 147–148].

While the above results confirm that a high mitigation productivity in developing countries can make both conditional monetary and in-kind transfers capable to bring about a Pareto improvement, this can, however, not be taken for granted as the discussion above has shown. For Case 2 of the matching scheme, transfers become welfare improving for both regions (see (29b) and (33b)), if the positive one of the two income effects in the developing country exceeds the negative substitution price effect and if the productivity of adaptation in generating additional units of the private characteristic is sufficiently high, i.e., if

$$\delta > \gamma + (1 - t\gamma)^2 p_D^{e,m} x_D \cdot \max\{A_D, A_I\} \tag{38}$$

with

$$A_D = \frac{\partial X_I}{\partial U_I}\left(x_D \frac{\partial X_I}{\partial U_I} + p_D^{e,m} \frac{\partial c_I}{\partial U_I}\frac{\partial X_D}{\partial p_D^{e,m}}\right)^{-1};$$

$$A_I = \frac{\partial X_D}{\partial U_D}\left(x_D \frac{\partial X_D}{\partial U_D} - \frac{\partial E_D}{\partial U_D}\frac{\partial X_D}{\partial p_D^{e,m}}\right)^{-1}.$$[28]

Assuming that the second-order effects of a reduction of $p_D^{e,m}$ are not too strong, the inequality in (38) is more likely to be fulfilled if $p_D^{e,m}$ is small. So, a high productivity differential between the regions makes a mutual profitability of matching more likely.[29] Yet, this alone cannot explain why international climate policy focuses so much on adaptation finance, as productivity differentials tend to affect the welfare of countries under both schemes (subsidy and matching) in the same direction.

5. Conclusions

Industrialized countries pledged USD 100 billion as an annual climate-related transfer towards the developing world by 2020. Our analysis investigated in which way such transfers channelled from industrialized to developing countries could lead to welfare improvements not only in the developing world but also in developed countries. As international spillover effects are a very important pay-back channel for industrialized countries' investments in transfers,

[28]For the subsidy scheme and Case 1 of the matching scheme, equivalent conditions under which both regions gain can be derived.

[29]Please note that matching can never be welfare enhancing for both regions if the positive income effect in the developing country is lower than the negative substitution price effect. In this case $\gamma+(1-t\gamma)^2 p_D^{e,m} A_I < \delta < \gamma+(1-t\gamma)^2 p_D^{e,m} A_D$ would have to hold. Yet, as $A_D < 0$ if the income effect is smaller than the substitution price effect, this condition can never be fulfilled.

we focused on schemes involving conditional transfers that reduce the effective price of the provision of the global public good "climate change mitigation" in developing countries. In one of these schemes (subsidy scheme), such mitigation efforts in the developing region were plainly subsidized via conditional monetary transfers. In an alternative scheme (matching scheme), mitigation efforts in the developing region were matched by conditional in-kind transfers dedicated to adaptation to climate change in developing countries.

By using an expenditure minimization approach, we confirmed that conditional transfers may harm the transfer receiving region while they tend to improve the welfare of the donor region. This seemingly paradoxical result is already known from earlier studies (e.g., Bergstrom [1989] and Ihori [1996]). We, however, also showed that it is possible for both regions to win in the case of conditional adaptation transfers, a result which is — to our knowledge — new to the literature. Whether regions win or lose depends crucially on the feedback effect of adaptation on the benefits from mitigation in the developing region. Not surprisingly, we also found that reducing effective prices of climate change mitigation via conditional transfers (both matching and subsidies) will bring about a higher level of climate change mitigation in the transfer receiving region.

From the developing countries' perspective, whether or not adaptation support (matching) would be preferred to a subsidy scheme, depends not only on the productivity of adaptation but also on the relative strength of income and substitution price effects. In contrast to a subsidy scheme, matching can result in a welfare increase for developing countries even if they do not possess a comparative advantage with respect to their mitigation productivity. Yet, even if adaptation technologies were favorable in the sense that such a positive welfare effect could theoretically arise, setting the appropriate transfer rate would require knowledge about the strength of the substitution and income effects which is rather unrealistic.

From the donors' point of view, a direct comparison of subsidy and matching schemes shows that matching tends to be less attractive than subsidization if adaptation and mitigation are strong substitutes. This tends to hold regardless of the ex-ante productivity of mitigation as reflected by the effective pre-policy price of the public good. However, a low pre-policy price in the developing region makes the absolute profitability of matching (and subsidization) more likely. Without this absolute profitability, i.e., without beneficial or with adverse effects, agents would not choose matching although it is more attractive (less adverse) than subsidization.

The pending relative disadvantage of the matching scheme from the donor's point of view stems from the adverse effect of adaptation support on the productivity of mitigation in developing countries, which tends to cause a decline in these countries' willingness to contribute to global public good provision. Therefore, even when adaptation support schemes may be beneficial for both considered regions, the subsidy scheme may remain superior from the donors' perspective. Consequently, from the donors' point of view, self-protection via domestic adaptation to climate change might be preferable to adaptation support in developing countries. According to Zehaie [2009: 349], this domestic "self-protection has strategic advantages as it transfers greater responsibility for international environmental problems to other countries." Support of adaptation abroad, however, implies an export of such advantages to the transfer receiving countries.

We found, however, also that — depending on the calibration of the model — matching can become more attractive than subsidization from the industrialized region's point of view. We showed that this outcome becomes more likely with low adverse mitigation productivity effects of adaptation and high transfer rates. It is especially the adaptation technology's productivity that can make matching more attractive and may therefore — from an allocation point of view — justify the current high interest in the climate policy arena to provide adaptation support: The matching scheme might become the preferred scheme from both transfer donors' and recipients' point of view, when productivity of the adaptation technology provided by the industrialized region in raising private consumption is very high.

Furthermore, we have seen that situations may arise where developing countries might even have incentives to misguide or waste adaptation support. Decision makers should be aware of the risk that principal-agent situations may arise where the agents receiving transfers have an interest to misguide funds in order to gerrymander the prevailing scheme such that their domestic welfare is maximized. Incentives should be given to discourage transfer receiving regions from doing so.

Summing up, our above findings show that while an allocation-based justification of international adaptation transfers as substitutes for mitigation subsidies can be found, this holds, even in our highly stylized and simplified world, only under very specific circumstances. For an evaluation of whether or not these circumstances reflect reality, empirical evidence and thus more research on the parameters of the model — especially δ and γ — would be required.

Matters become even more complex if one considers intertemporal aspects of adaptation support. As adaptation support might only be superior to a subsidy scheme in bringing about a welfare improvement for both involved regions if the adaptation productivity remains permanently high (and even then, the success is not secure), questions about the adequate timeframe for international (adaptation) support have to be raised. Even if the USD 100 billion pledge (implying also large amounts of funding dedicated to adaptation) were advantageous in the presence and near future, this does not necessarily hold in the mid-term. As wealth and also the technological prospects (for adaptation) of individual developing countries improve, the beneficial allocative consequences of channeling international transfers towards them may deteriorate. Hence, a time-limit for or periodic reevaluation of the transfer-pledge from developed countries seems advisable.

Finally, the lack of a stipulated conditionality of adaptation support in the Cancun Accords (i.e., adaptation support will not explicitly be subjected to marginal increases in developing regions' mitigation efforts) further raises skepticism concerning the profitability of adaptation transfers from the donors' point of view. However, implicitly — as the specific requirements for international funds to be accounted for the USD 100 billion pledge are not yet agreed upon — there might be some conditionality pending. If the developing countries would be defective in establishing a future global climate agreement and refuse to make own abatement commitments, industrialized countries might, e.g., define "additionality" of climate funds in a rather broad way. As some recent calculations already suggest (see e.g., Buchner *et al.* [2011]) that current international climate finance almost achieves the USD 100 billion mark, and thus, depending on the industrialized world's re-labeling efforts concerning these funds, the amount of new international financial flows dedicated to climate policy would vary.

From our results, we conclude that the theoretical evidence for an allocation-based motivation for international adaptation transfers remains mixed at best. Reasons for adaptation transfers might thus mainly be found in the "distributional branch" [Musgrave, 1959] of public policy, i.e., equity (see, e.g., Mendelsohn [2012] who makes a distinction between developed, developing and emerging countries) and fairness (see, e.g., Rübbelke [2011], Pittel and Rübbelke [2013]) considerations may be important. Aspects concerning international law may also play a role (see, e.g., Tol and Verheyen [2004]) as there might be the possibility that countries are held liable under international law if their emissions of greenhouse gases were to cause damage to other countries. Major emitters might seek to lower such a liability risk by offering adaptation assistance to strongly affected countries which in turn tends to reduce the climate change damage.[30]

References

Akai, N. 2003. When Do Cost Differentials among Privately Provided Public Goods Make Income Transfer Policy Effective?. *Economics Bulletin* 8, 1–7.

Althammer, W. and W. Buchholz. 1993. Lindahl-Equilibria as the Outcome of a Non-Cooperative Game. *European Journal of Political Economy* 9, 399–405.

[30]There are currently strong international disputes about how to address "loss and damage" associated with climate change impacts. At the COP in Doha in 2012, the first time the phrase "loss and damage" had been explicitly addressed in an international legal document [UNFCCC, 2012].

Auerswald, H., K. A. Konrad and M. Thum. 2011. *Adaptation, Mitigation and Risk-Taking in Climate Policy.* Working Paper No. 3320. CESifo. Munich, Germany.

Barrett, S. 2008, Climate Treaties and the Imperative of Enforcement. *Oxford Review of Economic Policy* 24, 239–258.

Bergstrom, T. C. 1989. Puzzles — Love and Spaghetti, The Opportunity Cost of Virtue. *Journal of Economic Perspectives* 3, 165–173.

Bosello, F., C. Carraro and E. de Cian. 2010. Climate Policy and the Optimal Balance between Mitigation, Adaptation and Unavoided Damage. *Climate Change Economics* 1, 71–92.

Buchholz, W., R. C. Cornes and D. Rübbelke. 2011. Interior Matching Equilibria in a Public Good Economy: An Aggregative Game Approach. *Journal of Public Economics* 95, 639–645.

Buchholz, W. and K. A. Konrad. 1994. Global Environmental Problems and the Strategic Choice of Technology. *Journal of Economics* 60, 299–321.

Buchholz, W. and K. A. Konrad. 1995. Strategic Transfers and Private Provision of Public Goods. *Journal of Public Economics* 57, 489–505.

Buchholz, W. and W. Peters. 2005. A Rawlsian Approach to International Cooperation. *Kyklos* 58, 25–44.

Buchner, B., A. Falconer, M. Hervé-Mignucci, C. Trabacchi and M. Brinkman. 2011. *The Landscape of Climate Finance.* CPI Report. Climate Policy Initiative. Venice, Italy.

Buob, S. and G. Stephan. 2011a. *On the Incentive Compatibility of Funding Adaptation.* NCCR Research Paper 2011/02. University of Bern.

Buob, S. and G. Stephan. 2011b. To Mitigate or to Adapt: How to Confront Global Climate Change. *European Journal of Political Economy* 27, 1–16.

Cornes, R. C. 1992. *Duality and Modern Economics.* Cambridge, UK: Cambridge University Press.

Cornes, R. C. and R. Hartley. 2007. Aggregative Public Good Games. *Journal of Public Economic Theory* 9, 201–219.

Cornes, R. C. and T. Sandler. 1984. Easy Riders, Joint Production and Public Goods. *Economic Journal* 94, 580–598.

Cornes, R. C. and T. Sandler. 1994. The Comparative Static Properties of the Impure Public Good Model. *Journal of Public Economics* 54, 403–421.

Cornes, R. C. and T. Sandler. 1996, *The Theory of Externalities, Public Goods and Club Goods.* Cambridge, UK: Cambridge University Press.

Danziger, L. and A. Schnytzer. 1991. Implementing the Lindahl Voluntary-Exchange Mechanism. *European Journal of Political Economy* 7, 55–64.

De Bruin, K. C., R. B. Dellink and R. S. J. Tol. 2009. AD-DICE: An Implementation of Adaptation in the DICE Model. *Climatic Change* 95, 63–81.

Diamond, P. A. and D. L. McFadden. 1974. Some Uses of the Expenditure Function in Public Finance. *Journal of Public Economics* 3, 3–21.

Ebert, U. and H. Welsch. 2012. Adaptation and Mitigation in Global Pollution Problems: Economic Impacts of Productivity, Sensitivity, and Adaptive Capacity. *Environmental and Resource Economics* 52, 49–64.

Ekins, P. 1996. How Large a Carbon Tax Is Justified by the Secondary Benefits of CO_2 Abatement?. *Resource and Energy Economics* 18, 161–187.

Gorman, W. M. 1980. A Possible Procedure for Analysing Quality Differentials in the Egg Market. *Review of Economic Studies* 47, 843–856.

Guttman, J. M. 1978. Understanding Collective Action: Matching Behavior. *American Economic Review* 68, 251–255.

Guttman, J. M. 1987. "A Non-Cournot Model of Voluntary Collective Action. *Economica* 54, 1–19.

Heuson, C., W. Peters, R. Schwarze and A.-K. Topp. 2012. *Which Mode of Funding Developing Countries' Climate Policies under the Post-Kyoto Framework?*, Discussion Paper 10/2012. UFZ. Leipzig, Germany.

Ihori, T. 1996. International Public Goods and Contribution Productivity Differentials. *Journal of Public Economics* 61, 139–154.

Ingham, A., J. Ma and A. Ulph. 2005a, *Can Adaptation and Mitigation Be Complements?*, Working Paper 79. Tyndall Centre for Climate Change Research. Norwich, UK.

Ingham, A., J. Ma and A. Ulph. 2005b, *How Do the Costs of Adaptation Affect Optimal Mitigation When There Is Uncertainty, Irreversibility and Learning?*. Working Paper 74. Tyndall Centre for Climate Change Research. Norwich, UK.

Ingham, A., J. Ma and A. Ulph. 2007. Climate Change, Mitigation and Adaptation with Uncertainty and Learning. *Energy Policy* 35, 5354–5369.

Kane, S. and J. F. Shogren. 2000. Linking Adaptation and Mitigation in Climate Change Policy. *Climatic Change* 45, 75–102.

Lancaster, K. J. 1966, A New Approach to Consumer Theory. *Journal of Political Economy* 74, 132–157.

Mendelsohn, R. 2012. The Economics of Adaptation to Climate Change in Developing Countries. *Climate Change Economics* 3, 21 pages.

Musgrave, R. A. 1959. *The Theory of Public Finance.* New York, NY: McGraw Hill.

Neary, J.P. and K. W. S. Roberts. 1980. The Theory of Household Behaviour under Rationing. *European Economic Review* 13, 25–42.

Ono, T. 1998, Consumption Externalities and the Effects of International Income Transfers on the Global Environment, *Journal of Economics* 68, 255–269.

Paavola, J. and W.N. Adger. 2006. Fair Adaptation to Climate Change. *Ecological Economics* 56, 594–609.

Pittel, K. and D. T. G. Rübbelke. 2013. International Climate Finance and Its Influence on Fairness and Policy. *The World Economy* 36, 419–436.

Pittel, K. and D. T. G. Rübbelke. 2015. On Fostering International Public Good Provision: Would Complementarity between Public Good and In-Kind Transfers Help?. *Economics Bulletin* 35 (3), 1638–1644.

Rothbarth, E. 1940–41. The Measurement of Changes in Real Income under Conditions of Rationing. *Review of Economic Studies* 8, 100–107.

Rübbelke, D. T. G. 2002. *International Climate Policy to Combat Global Warming: An Analysis of the Ancillary Benefits of Reducing Carbon Emissions.* Cheltenham, UK: Edward Elgar.

Rübbelke, D. T. G. 2011. International Support of Climate Change Policies in Developing Countries: Strategic, Moral and Fairness Aspects. *Ecological Economics* 70, 1470–1480.

Schenker, O. 2013. Exchanging Goods and Damages: The Role of Trade on the Distribution of Climate Change Costs. *Environmental and Resource Economics* 54, 261–282.

Schenker, O. and G. Stephan. 2012. *Terms-of-Trade and the Funding of Adaptation to Climate Change and Variability: An Empirical Analysis.* Discussion Paper No. 12-056. ZEW. Mannheim, Germany.

Tol, R. S. J. 2005. Adaptation and Mitigation: Trade-offs in Substance and Methods. *Environmental Science and Policy* 8, 572–578.

Tol, R. S. J. and R. Verheyen. 2004. State Responsibility and Compensation for Climate Change Damages — A Legal and Economic Assessment. *Energy Policy* 32, 1109–1130.

United Nations Framework Convention on Climate Change (UNFCCC). 2011. *Report of the Conference of the Parties on its Sixteenth Session*; *Addendum: Part Two: Action Taken by the Conference of the Parties at its Sixteenth Session.* FCCC/CP/2010/7/Add.1.

United Nations Framework Convention on Climate Change (UNFCCC). 2012. *Report of the Conference of the Parties on its Eighteenth Session, Held in Doha from 26 November to 8 December 2012, Addendum: Part Two: Action Taken by the Conference of the Parties at Its Eighteenth Session* [Internet]; [updated 2016 January 22]. At: http://unfccc.int/resource/docs/2012/cop18/eng/08a01.pdf.

United Nations Framework Convention on Climate Change (UNFCCC). 2015. *Paris Agreement* [Internet]; [updated 2016 January 22]. At: https://sustainabledevelopment.un.org/frameworks/parisagreement.

Zehaie, F. 2009. The Timing and Strategic Role of Self-Protection. *Environmental and Resource Economics* 44, 337–350.

Chapter 12

Technological Transfers in Global Climate Policy — A Strategic Perspective

Wolfgang Buchholz

University of Regensburg,
Department of Economics and CESifo Munich,
93040 Regensburg, Germany
Wolfgang.Buchholz@ur.de

Lisa Dippl

University of Regensburg, Department of Economics,
93040 Regensburg, Germany

Michael Eichenseer

University of Regensburg, Department of Economics,
93040 Regensburg, Germany

Abstract

Pioneering activities in climate policy through increased abatement efforts or financial transfers are costly but often not very successful. In this chapter, we show that leading behavior by a coalition of countries, which is instead based on green technological innovations by members of the coalition and a subsequent costless transfer of the improved technology to other countries, may provide a sensible approach for combatting global warming through unilateral action. In particular, we examine in an otherwise standard model of private public good supply how the effects of such a policy on total greenhouse gas mitigation and welfare of countries depend on the level of research and development (R&D) costs, the intensity and scope of technological spillovers, the size of the technological coalition and the number of countries outside the coalition. In particular, it is shown that such technological transfers, which can be considered as an instrument of "climate finance" in a broader sense, may be required to induce in the strategic context of global public good provision R&D-activities and thus leading to the formation of a technological coalition.

1. Introduction

Progress in global climate policy is impeded by the enormous difficulties to ensure international cooperation and coordination on greenhouse gas mitigation, i.e., to conclude an international climate agreement, which features ambitious and effectively binding obligations and which is stable [see, e.g., Finus, 2001; Sandler, 2004]. The hope therefore is that — following a "bottom-up approach" — pioneering activities of a coalition of leading countries might help moderate global warming and thus increase the supply of this presumably most important public good. When such leading activities are considered, the focus normally is on increased abatement efforts or transfers to outsider countries. But such unilateral measures are only of limited effectiveness as the countries outside the coalition can be expected to show some "crowding-out" behavior, i.e., to decrease their public good contributions as a reaction to the increased contributions made by the coalition so that, in the case of climate change, "carbon leakage" results. In Sinn's [2012] figurative terminology, these outsider countries thus are "grabbing from the collection box" which reduces the positive effect on public good supply resulting from the pioneering activities [see, e.g., Andreoni, 1988; Buchholz *et al.*, 1998]. Even worse, it is even possible that total public good supply falls when countries unilaterally raise their abatement efforts before they enter climate negotiations [see Hoel, 1991].

In order to cope with these unfavourable conditions for successful leadership[1] and to make climate policy as a whole more effective,

[1] A detailed analysis of various determinants for successful leadership in climate policy is given by Buchholz and Sandler [2017] where, in particular, aspects of

a crucial role is attributed to "climate finance", i.e., international transfers which should help poor countries to undertake additional mitigation and adaptation measures and which in general should promote their willingness to participate in global climate cooperation.[2] The different forms and channels of climate finance have been explored in a lot of empirical [see Buchner *et al.*, 2011; Stavins and Ji, 2014, pp. 1038–1044, for overviews] and theoretical contributions, where it has been shown how both unconditional transfers [see Buchholz and Konrad, 1995] and conditional "matching" transfers [see, e.g., Buchholz *et al.*, 2015, and the literature cited there] can enhance efficiency of public good provision and which incentive problems are to be observed in this context.[3] It has also been known for a long time that international transfers may serve as an instrument to stabilize climate coalitions [see already Carraro and Siniscalco, 1993, and more recently the survey by Carraro *et al.*, 2006].

Instead of financial transfers and matching contributions, this chapter considers another type of transfers, which also aims at improving global climate policy: Developing "green" technologies (such as renewable energies or energy efficiency measures) may help reduce abatement costs not only in the countries where the research and development (R&D) efforts are made but also in other countries which get access to the improved climate-friendly technologies.[4] Dissemination of technology thus works like some transfer in kind and can thus be perceived as a substitute or equivalent for "climate finance" in the narrow sense.

But similar as in the matching scenario, green R&D is facing the problem that the strategic context of global public good provision

---

behavioral economics are integrated into the model of voluntary public good provision.

[2]Note that under certain conditions financial transfers may have the same effects on the allocation as increased public good contributions [see, e.g., Cornes and Sandler, 2000].

[3]A general discussion of climate finance especially referring to its ethical aspects is provided by Rübbelke [2011].

[4]The reasons for and mechanisms and channels of technology transfers in the case of climate change are addressed by Newell [2010], Popp [2011, 2014] and Stavins and Ji [2014, pp. 1035–1037].

may undermine the incentives for innovation. So it is well-known [see Buchholz and Konrad, 1994; Ihori, 1996] that it may harm a country if it unilaterally reduces its cost for public good provision even in the extreme case where the innovation is completely costless. The reason for this paradoxical effect is that a country-specific cost reduction simultaneously stimulates the willingness of the innovating country to increase its public good contribution which in turn leads to a crowding-out of the other countries' contributions in the non-cooperative solution and weakens the bargaining position of the technological forerunner in subsequent international climate negotiations [see, e.g., Tirole, 2012]. Then environmentally friendly technological progress is blocked on the basis of strategic reasons.

In the literature on international cooperation, the development of climate-friendly technology has only been dealt with incidentally until now. The focus in the existing work instead is on the total welfare effects of technological cooperation, which, e.g., depend on whether carbon taxes or emission trading are applied as instruments of climate policy [see Golombek and Hoel, 2005], and specifically on the stability of the technology and abatement coalitions [see, e.g., Barrett, 2006; Helm and Schmidt, 2014; Hoel and de Zeuw, 2010].

In this chapter, we extend the analysis of the strategic choice of abatement technology[5] in a two-stage setting by exploring the incentives for doing green R&D in a coalition of willing countries and for formation of such an innovating coalition at all. This leads us to a more optimistic picture: If a coalition of countries makes the fruits of its innovation available to other countries free of charge this non-financial transfer may be beneficial for the pioneering countries, which in turn makes the coalition more prepared to engage in green R&D-activities. An incentive problem, however, arises on the side of the countries for which the innovative technology is made available since these countries might do better by not adopting it.

[5]In a different framework, strategic technology development in climate change policy is explored by Urpelainen [2013]. In a repeated game, Harstad *et al.* [2015] consider a strategic effect of the choice of the abatement technology through which countries can improve the credibility of their future cooperation.

In our analysis we assume that there is some kind of core group of countries, which is both prepared and able to conduct "green" R&D. The problem when a technology coalition is stable therefore will only be briefly addressed in this chapter.

The analysis will be carried out in the standard framework of voluntary public good provision [as devised by Bergstrom, Blume and Varian, 1986; Cornes and Sandler, 1996]. But unlike these traditional contributions to the theory of public goods we will make use of the more recent Aggregative Game Approach[6] [see Cornes and Hartley, 2007]. Application of this approach considerably facilitates the analysis of Nash equilibria in games of public good provision, which is quite complex in the scenario considered in this chapter.

The paper is organized as follows: In Section 2 we present the public good model, in which our analysis is conducted. In Section 3 we describe the technological interdependencies between three groups of countries, i.e., the innovating coalition, the group of countries which adopt the improved mitigation technology, and the group of complete outsiders which keep the old technology. Section 4 presents some comparative statics focussing on the change of public good supply, which is implied by the innovation and the subsequent technology transfer. How utility of the coalition members is changed by the innovation is explored in Section 5, while Section 6 deals with the welfare effects that the innovation has on the members of the other two groups and the ensuing incentive problem. After illustrating the general results of our chapter through a Cobb-Douglas example in Section 7, we conclude in Section 8 where we discuss a few empirical applications of our analysis.

2. The Framework

Let there be n countries $i = 1, \dots, n$, which all have the same initial endowment w of the private good and the same utility function $u(x_i, G)$ where x_i denotes private consumption of country i and G

[6]Specific applications of the Aggregative Game Approach to environmental economics are presented in Cornes [2016].

is public good supply. Especially in the context of climate change G can be interpreted as the amount of greenhouse gases totally avoided. The utility function is assumed to have the standard properties, i.e., it is twice partially differentiable with the first partial derivatives $u_1(x_1, G) > 0$ and $u_2(x_i, G) > 0$ and quasi-concave. Moreover, we suppose that the private and the public good both are strictly non-inferior, i.e., demand for both of them is increasing when the initial endowment grows while the rate of transformation between the private and the public good is fixed.

The crucial element for our analysis is that the countries may differ in their productivities for generating the public good which are represented by the country-specific marginal rates of transformation mrt_i between the private and the public good, which is denoted by a_i: These productivities indicate how many units of the public good country i can produce if it spends one unit of the private good for public good provision. The reciprocal value $c_i = \frac{1}{a_i}$ then indicates how many units of the public good country i has to give up in order to get one additional unit of the private good. In the case of environmental public goods c_i thus represents the marginal abatement costs of country i.

Under these assumptions a feasible allocation $(x_1, \ldots, x_n, G)$ has to satisfy the aggregate budget constraint

$$G = \sum_{i=1}^{n} a_i(w - x_i) \tag{1}$$

or, equivalently,

$$G + \sum_{i=1}^{n} a_i x_i = \sum_{i=1}^{n} a_i w. \tag{2}$$

In order to specify a country's possible equilibrium positions in a Nash equilibrium of voluntary public good provision, let $e(G, \alpha_i)$ be country i's (income) expansion when α_i is the marginal rate of substitution mrs_i between the private and the public good. As non-inferiority of both goods is assumed, these expansion paths are well-defined and strictly monotone increasing in G as depicted in Figure 1.

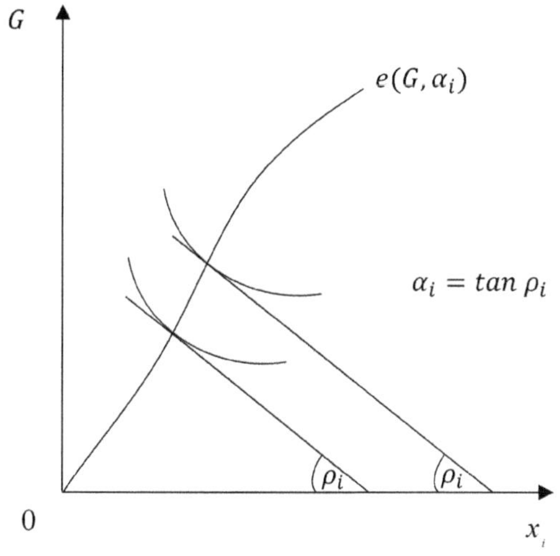

Figure 1. An Income Expansion Path.

Along $e(G, \alpha_i)$, the indifference curves of a country all have the same slope $-\alpha_i = -\tan\rho_i$. As an additional assumption regarding preferences, we assume that $e(0, \alpha_i) = 0$ and $\lim_{G\to\infty} e(G, \alpha_i) = \infty$, which means that demand for the public good does not fall to zero for a finite value of initial endowment and that demand for the private good is not bounded above when the initial endowment goes to infinity.

The essential point for the characterization of Nash equilibria now is that a country, which makes a strictly positive public good contribution, is in an individual equilibrium position only if its mrs_i coincides with its mrt_i, i.e., $\alpha_i = a_i$ holds. Otherwise, country i could attain a higher utility level either by slightly increasing (if $\alpha_i > a_i$) or by slightly decreasing (if $\alpha_i < a_i$) its public good contribution. This, however, means that in an interior Nash equilibrium $(\hat{x}_1, \ldots, \hat{x}_n, \hat{G})$ with positive public good contributions of all countries, country i's position $(\hat{x}_i, \hat{G})$ has to lie on the expansion path $e(G, a_i)$ such that $\hat{x}_i = e(\hat{G}, a_i)$.

As a benchmark for our analysis, we consider the case in which the productivity parameters a_i are fixed and no country undertakes efforts to improve public good productivity. Then no R&D-costs have to be taken into account, and the budget constraint (1) for an interior Nash equilibrium becomes

$$\hat{G} = \sum_{i=1}^{n} a_i(w - e(\hat{G}, a_i)). \tag{3}$$

Given our assumptions the function $\Phi(G) := \sum_{i=1}^{n} a_i(w - e(G, a_i))$, whose value at $\hat{G}$ appears on the right-hand side of equation (3), is strictly monotone decreasing and continuous and, given our assumptions on expansion paths, has $\Phi(0) = \sum_{i=1}^{n} a_i w$ and $\lim_{G\to\infty} \Phi(G) = -\infty$. Hence, by the intermediate value theorem there exists exactly one level of public good supply $\hat{G}$, which fulfils condition (3). If $e(\hat{G}, a_i) < w_i$ holds for each country $i = 1, \ldots, n$, the Nash equilibrium is interior with public good supply $\hat{G}$ and private consumption levels $\hat{x}_i = e(\hat{G}, a_i)$. In this Nash equilibrium, country $i = 1, \ldots, n$ spends $\hat{z}_i = w_i - \hat{x}_i$ on the public good thus inducing an increase of public good supply by $\hat{g}_i = a_i \hat{z}_i$.

3. Technological Interdependencies

We start from a situation in which all countries have the same productivity parameter a_0. Public good supply $\hat{G}(a_0)$ in the Nash equilibrium — which is clearly interior in this initial state of full symmetry and without any R&D-costs — then is given by

$$\hat{G}(a_0) = na_0(w - e(\hat{G}(a_0), a_0)). \tag{4}$$

Now the possibility arises that a subgroup of countries undertakes some R&D-efforts aimed at improving "green" technologies (such as better insulation of houses, renewable energies, smart grids and new methods for power storage) through which the reduction of carbon emissions becomes cheaper or, in other words, the productivity of the global public good climate protection is increased. Through intended or unintended technological spillovers, other countries may also benefit from these productivity enhancing effects even if they do not incur any of the costs associated with developing these ecologically friendly technologies.

For a precise description of this scenario, we divide the whole group of n countries into three subgroups K, L and M whose members are playing a two-stage game.

3.1. *Subgroup K consisting of k countries*

The members of subgroup K are forming a technological coalition, which is willing to play a pioneering role in climate policy by collectively promoting green innovations. In the framework of our model, this means that at the first stage of the game coalition K is able to choose an improved production technology for the public good which exhibits a higher public good productivity a than the original technology. Choosing some $a > a_0$, however, is not costless but results in R&D-costs of $c_k(a)$ for each country in coalition K. This cost function is assumed to be differentiable in a and has $c_k(a_0) = 0$. If, as in the case of basic research, R&D-costs can be divided equally among the members of the coalition $c_k(a)$ will — for any $a > 0$ — fall when k increases. However, if technological progress is based on learning-by-doing activities, which are carried out in each country of the coalition at an equal scale, then $c_k(a)$ will not be affected by the size of the coalition.

While the coalition cooperates at the innovation stage, the coalition members still act independently in the second stage of the game at which the coalition members decide on their contributions to the public good. This assumption reflects the notion that in climate policy cooperation on abatement levels is harder to achieve than technological cooperation.

3.2. *Subgroup L consisting of l countries*

The members of subgroup L do not have a share in the spillover: They stick to the original technology with the productivity parameter a_0 irrespective of the technological choice made by coalition K. This inability to make use of the better environmental technologies may arise from specific physical or meteorological conditions. E.g., countries in the tropical zones obviously do not benefit from improved efficiency in the heating of buildings, and countries like

Canada with fewer sunshine hours than Florida cannot gain much from the development of solar technology. But it is also possible that in developing countries the capacities for adopting the improved technologies are lacking. In contrast to the physical limitations, these obstacles can be removed, e.g., through education and the formation of human capital.

3.3. *Subgroup M consisting of m countries*

For subgroup M, there is a technological spillover from the technological innovations provided by coalition K so that the members of this group become more productive in generating the public good — but possibly to a different degree than as the coalition members. The differentiable function $b(a)$ describes which productivity parameter results in each country in M when the productivity parameter chosen in coalition K is a. This function measuring the intensity of the spillover effect is monotone increasing in a with $b(a_0) = a_0$. The normal case will be $b'(a) \leq 1$, which means that the countries in M benefit not more from the innovation than the countries in K. Nevertheless, situations are conceivable in which $b'(a) > 1$ holds so that the productivity increase for subgroup M is even larger than in K. An example for this might be solar energy, when in the countries of subgroup M solar radiation is stronger than in the countries of subgroup K.

Like the countries in coalition K, the countries in the outsider subgroups L and M determine their public good contributions non-cooperatively at the second stage of the game.

Applying the Aggregative Game Approach, it is now straightforward to describe the interior Nash equilibrium which results when coalition K has chosen some productivity parameter $a \geq a_0$ as we know that in the Nash equilibrium

- the position of all countries in K is on the expansion path $e(G, a)$;
- the position of all countries in L is on the expansion path $e(G, a_0)$;
- the position of all countries in M is on the expansion path $e(G, b(a))$.

Based on condition (3), public good supply $\hat{G}(a)$ in the Nash equilibrium if coalition K has chosen the productivity parameter a is characterized by the following equation:

$$\hat{G}(a) = ka(w - e(\hat{G}(a), a) - c_k(a)) + la_0(w - e(\hat{G}(a), a_0)) \\ + mb(a)(w - e(\hat{G}(a), b(a))). \quad (5)$$

Private consumption of the countries in subgroups K, L and M thus is $\hat{x}_k(a) = e(\hat{G}(a), a)$, $\hat{x}_L(a) = e(\hat{G}(a), a_0)$ and $\hat{x}_M(a) = e(\hat{G}(a), b(a))$, respectively. Note that in equation (5), it is taken into consideration that the members of group K do not spend the whole residual between income and private consumption for public good provision because they have to spend $c_k(a) > 0$ for R&D-efforts when choosing some $a > a_0$.

Since the initial Nash equilibrium is interior, it follows from a standard continuity argument that the Nash equilibrium will stay interior when the productivity parameter a chosen by coalition K is sufficiently close to a_0. The analysis to follow only considers these cases.

4. The Change in Public Good Supply through Technological Progress

Let the partial derivative of any expansion path $e(G, \alpha)$ w.r.t. public good supply G be denoted by $e_1(G, \alpha)$. This derivative describes how private consumption changes if one is moving along an expansion path. Analogously, $e_2(G, \alpha)$ is the partial derivative of the expansion path w.r.t. the marginal rate of substitution α. This derivative indicates the change of private consumption, which results when — for a given level of public good supply — the move is to another expansion path corresponding to a higher marginal rate of substitution. From the non-inferiority assumption on preferences, we have $e_1(G, \alpha) > 0$ and $e_2(G, \alpha) < 0$.

To calculate the effect on public good supply $\hat{G}'(a) = \frac{\partial \hat{G}}{\partial a}$, which is driven by a marginal change of its productivity parameter by coalition K, we first consider the total differential of equation (5) at some arbitrary a for which interiority holds which yields

$$\begin{aligned}\hat{G}'(a) = k(w - e(\hat{G}(a),a) - c_k(a)) - ka(e_1(\hat{G}(a),a)\hat{G}'(a) \\ + e_2(\hat{G}(a),a) + c_k'(a)) - la_0e_1(\hat{G}(a),a_0)\hat{G}'(a) \\ + mb'(a)(w - e(\hat{G}(a),b(a))) - mb(a)(e_1(\hat{G}(a),b(a))\hat{G}'(a) \\ + e_2(\hat{G}(a),b(a))b'(a)). \qquad (6)\end{aligned}$$

We now apply equation (6) to infer the effects on public good supply, which result from a marginal change of a starting from $a_0 = b(a_0)$. Without loss of generality we can assume $a_0 = 1$ and, to simplify notation, we use abbreviations as follows: $\hat{G}' = \hat{G}'(1)$, $\hat{z} = w - e(\hat{G}(1),1)$, $\kappa_k = c_k'(1)$, $\beta = b'(1)$, $\gamma_1 = e_1(\hat{G}(1),1)$ and $\gamma_2 = e_2(\hat{G}(1),1)$. Since $c_k(1) = 0$ by assumption, condition (6) then turns into

$$\hat{G}' = k(\hat{z} - \gamma_1\hat{G}' - \gamma_2 - \kappa_k) - l\gamma_1\hat{G}' + m(\beta\hat{z} - \gamma_1\hat{G}' - \gamma_2\beta). \qquad (7)$$

Solving (7) for $\hat{G}'$ and observing $k + l + m = n$ give the following result.

Proposition 1. *If coalition K marginally increases its productivity parameter a starting from the symmetric Nash equilibrium with $a_0 = 1$, then public good supply changes by*

$$\hat{G}' = \frac{(k + m\beta)(\hat{z} - \gamma_2) - k\kappa_k}{1 + n\gamma_1}. \qquad (8)$$

Public good supply hence increases if and only if

$$k\kappa_k < (k + m\beta)(\hat{z} - \gamma_2). \qquad (9)$$

Since $\gamma_2 < 0$, condition (9) directly shows that an increase in public good supply always results if the aggregate marginal costs for the technological improvement $k\kappa_k$ are not too high. A high spillover parameter β and a high public good contribution $\hat{z}$ in the original Nash equilibrium are also favorable for an increase of public good supply as both help to make the increase of public good productivity more effective.

If, however, the R&D-costs are sufficiently high, so that $k\kappa_k > (k + m\beta)(\hat{z} - \gamma_2)$ holds, public good supply is reduced by the innovation. The reason for this adverse effect is that due to the costly R&D-efforts coalition K's resources available for public good provision are eaten up while at the same time the spillover effect is too weak, either because only few countries are positively affected or the intensity of the spillover is small.

In addition, we can infer from conditions (8) and (9) how for a fixed total number of countries n the size of the different subgroups affects the change of public good supply. In this context, we first note that $\hat{z}$, γ_1 and γ_2 refer to the original fully symmetric Nash equilibrium and thus do not depend on k, l and m as long as the total number of countries $n = k + l + m$ is fixed.

Proposition 2. *Assume that $\hat{G}'$ is positive. Then $\hat{G}'$ is the larger*

- *the larger the coalition K is when aggregate marginal costs $k\kappa_k$ of the technological improvement are not rising in k;*
- *the larger the group M is;*
- *the smaller $\gamma_1 > 0$ and the larger $-\gamma_2 > 0$ are.*

In a Nash equilibrium, public good supply normally is too low as compared to Pareto optimal levels [see Buchholz and Peters, 2001, for a treatment especially of exceptions]. Against this background, Proposition 2 says that this "underprovision" is mitigated both through a spatial expansion of the technological spillover, i.e., an increase of m, and an increase of its intensity β. The same positive effect on public good supply occurs if the coalition K is enlarged given that $\beta \leq 1$ and $k\kappa_k$ is decreasing in k.

For a further interpretation of Proposition 2, note that a small γ_1 means that in a x_i-G-diagram, in which the position of a representative country i is depicted, the expansion path $e(G, 1)$ is relatively steep. Then along this expansion path an increase of public good supply (from $\hat{G}(1)$ to $\hat{G}(a)$ for some $a > 1$) is accompanied only by a small increase in private consumption, so that the public good contribution of country i is not reduced so much and the crowding-out effect is small. If, on the other hand, $-\gamma_2$ is large the increase of

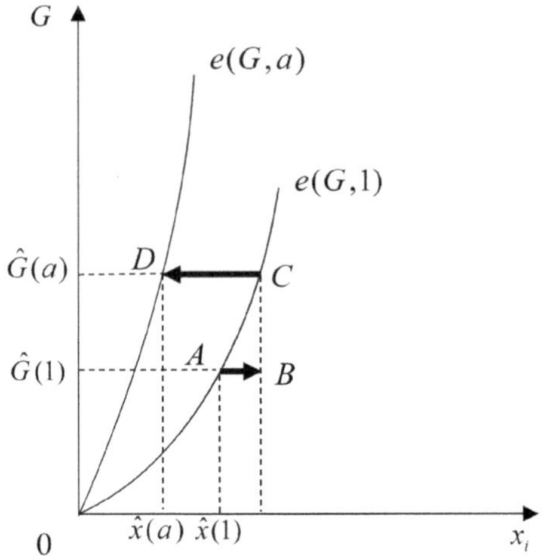

Figure 2. Moving Along an Expansion Path and Shifting the Expansion Path.

a induces a strong shift of the expansion path to the left and thus causing a large increase of country i's public good contribution. In Figure 2 the first (negative) effect on country i's public good contribution, which is described by the arrow from point A to point B, is dominated by the second (positive) effect, which is described by the arrow from C to D. The total effect on country i's public good contribution thus is positive. This is described by the shift from A to D in Figure 2.

As a next step, we examine the incentives the coalition K has for making a green innovation through which its public good productivity is increased.

5. The Incentives for Coalition K to Make the Technological Improvement

Given some productivity parameter a utility of a member of coalition K is $\hat{u}_k = u(e(\hat{G}(a), a), \hat{G}(a))$ in the Nash equilibrium as $e(\hat{G}(a), a) = \hat{x}_k(a)$ is its private consumption. A marginal variation of a changes this utility by

$$\hat{u}'_k(a) = u_1(\hat{x}_k(a), \hat{G}(a))(e_1(\hat{G}(a), a)\hat{G}'(a) + e_2(\hat{G}(a), a)) + u_2(\hat{x}_k(a), \hat{G}(a))\hat{G}'(a). \qquad (10)$$

Without loss of generality, we can assume that at the original Nash equilibrium for $a_0 = 1$ we have $\hat{u}_1(\hat{x}_k(1), \hat{G}(1)) = \hat{u}_2(\hat{x}_k(1), \hat{G}(1)) = 1$. With the abbreviations as introduced before and additionally letting $\hat{u}'_k = \hat{u}'_k(1)$, equation (10) is then reduced to

$$\hat{u}'_k = (1 + \gamma_1)\hat{G}' + \gamma_2. \qquad (11)$$

Based on equation (11), a precise condition for an increase of utility for countries in the coalition K is provided by the next result. In its first part, this Proposition is a direct consequence of equation (11) and in its second part it follows from plugging $\hat{G}'$ as given by equation (8) into equation (11).

Proposition 3. *Starting from the Nash equilibrium with $a_0 = 1$, the members of coalition K benefit from an increase of their public good productivity if and only if*

$$\hat{G}' > \frac{-\gamma_2}{1 + \gamma_1} > 0 \qquad (12)$$

holds or, equivalently, if and only if

$$k\kappa_k < (k + m\beta)(\hat{z} - \gamma_2) + \gamma_2 \frac{1 + n\gamma_1}{1 + \gamma_1}. \qquad (13)$$

As $\gamma_1 > 0$ and $\gamma_2 < 0$, it follows from condition (12) that a higher public good supply is a necessary but not a sufficient condition for an increase of a coalition member's utility: The coalition members only benefit from their R&D-efforts when the increase in public good supply is strong enough. Only if conditions (12) and (13) are satisfied the countries in K are willing to form a technological coalition in the first place.

The factors which determine the right-hand side of inequality (13) clearly are similar to those characterizing the change of public good supply: An enlargement both of the coalition K and of the group M are favorable for an increase of utility for the members of K.

Concerning the incentives for innovation in K, this in particular shows how important it is to ensure a broad dissemination of the improved technologies. Giving patents for green technological innovations to other countries away free of charge thus may be a clever strategic move for coalition K — and enhances the prospects that a R&D-coalition is forming at all.

The crucial role that access to climate-friendly technologies may play for climate policy has already been taken into account by the United Nations Framework Convention on Climate Change in 1992 where it has been stated in Article 4(5) that the developed countries "shall take all practical steps to promote, facilitate, and finance... the transfer of, or access to, environmentally sound technologies and know-how to other Parties, in particularly developing country Parties ...." Our analysis has shown that such a transfer may also lie in the vested interest of the developed countries.[7]

Concerning the second term on the right-hand side of (13), we note that $\frac{1+n\gamma_1}{1+\gamma_1}$ is increasing in γ_1. Hence, a utility increase for countries in coalition K is more likely if γ_1 is small. The effect of γ_2, however, is ambiguous.

In our analysis it had been assumed until now that the size of the coalition is exogenously given. But in a little different scenario inequality (13) can also be used to determine the size of an internally stable coalition endogenously which links our analysis to the main strand of the literature on technological coalitions in climate policy (see, e.g., Barrett, 2006; Helm and Schmidt, 2014; Hoel and de Zeuw, 2010). For a brief explanation look at the case in which group L is not present so that $m = n - k$. Further assume that (13) holds for $k = n$ but not for $k = 1$, and let $k^* \geq 2$ denote the minimum integer for which (13) is valid. Then a technological coalition of size k^*, which increases its productivity parameter to some a slightly above $a_0 = 1$,

[7]General arguments for a broad dissemination of newly created knowledge and thus against too restrictive patent laws are provided by Greenwald and Stiglitz [2014, Chapter 12].

is clearly internally stable since the remaining coalition of size $k^* - 1$ is not willing anymore to conduct the technological improvement if one country breaks off. Consequently, productivity remains at its original level $a_0 = 1$, and the deviating country gets worse by leaving the coalition.

6. Utility Effects for the Countries outside the Coalition

We now examine how utility of the countries in the outsider groups L and M is changed by the innovative activities of coalition K.

Differentiating utility $\hat{u}_L(a) = u(\hat{x}_L(a), \hat{G}(a)) = u(e(\hat{G}(a), a_0), \hat{G}(a))$ of a country in L and utility $\hat{u}_M(a) = u(\hat{x}_M(a), \hat{G}(a)) = u(e(\hat{G}(a), b(a)), \hat{G}(a))$ of a country in group M w.r.t. the productivity parameter a yields

$$\hat{u}'_L(a) = u_1(\hat{x}_L(a), \hat{G}(a))e_1(\hat{G}(a), a_0)\hat{G}'(a) + u_2(\hat{x}_L(a), \hat{G}(a))\hat{G}'(a) \tag{14}$$

$$\begin{aligned}\hat{u}'_M(a) = {} & u_1(\hat{x}_M(a), \hat{G}(a))(e_1(\hat{G}(a), b(a))\hat{G}'(a) + e_2(\hat{G}(a), b(a))b'(a)) \\ & + \hat{u}_2(\hat{x}_M(a), \hat{G}(a))\hat{G}'(a).\end{aligned} \tag{15}$$

Assuming again $a_0 = 1$ and $u_1(\hat{x}(1), \hat{G}(1)) = u_2(\hat{x}(1), \hat{G}(1)) = 1$ and abbreviating $\hat{u}'_L = \hat{u}'_L(1)$ and $\hat{u}'_M = \hat{u}'_M(1)$ a marginal change of productivity at the initial Nash equilibrium thus results in utility changes as follows:

$$\hat{u}'_L = (1 + \gamma_1)\hat{G}' \tag{16}$$

$$\hat{u}'_M = (1 + \gamma_1)\hat{G}' + \beta\gamma_2. \tag{17}$$

Comparing the utility changes for the three groups K, L and M as described by equations (11), (16) and (17) leads to the following result:

Proposition 4. *If coalition K marginally increases its public good productivity starting from the Nash equilibrium with $a_0 = 1$, the countries in K benefit least while countries in the group L benefit most, i.e., $\hat{u}'_k \leq \hat{u}'_M \leq \hat{u}'_L$.*

The interpretation of Proposition 4 is as follows: Through the change of public good productivity in coalition K, utility of countries in each subgroup is equally affected by$(1 + \gamma_1)\hat{G}'$, which is positive if public good supply increases. For countries in K there is, however, a negative partial effect on utility which is expressed by $\gamma_2 < 0$ and which reflects the increased willingness to pay for the public good when productivity improves. The same effect hits the group M but to a lesser degree if the spillover is incomplete, i.e., $\beta < 1$. If, however, $\beta = 1$ the utility change is the same for group K and group M even though only the members of the coalition K initially bear the cost of the green innovation. This means that, due to equilibrium repercussions, R&D-costs can be shifted to other countries. This indirect redistribution effect is, in a certain sense, similar to the famous Warr neutrality in voluntary public good provision [see Warr, 1982, and e.g., Cornes and Sandler, 1996], which in particular implies that in an interior Nash equilibrium an increase of income in some country will increase utility not only in that specific country but in all countries.

The negative effect, which arises from the change of the willingness to pay for the public good implied by the technological improvement, is completely absent for countries in group L whose technology is unaffected by the innovation. Therefore, the members of this group benefit most. This, however, creates an incentive problem because countries in group M would attain a higher utility level if they did not adopt the better technology for public good provision by becoming a member of group L. This strategic effect, however, is avoided if the technological spillover occurs automatically which, e.g., is the case if firms in coalition K are the dominant producers of energy technology and thus can set environmentally friendly standards worldwide [see, e.g., Barrett, 2003].

For the countries in group M also co-benefits from climate friendly technology may arise which, on the one hand, may be caused by improved possibilities to abate locally damaging pollutants as, e.g., particulate matter from power plants [see, e.g., Finus and Rübbelke, 2013] and, on the other hand, by the prospect of initiating a sustainable growth process implied by the transition to a low-carbon economy [see Stern, 2015]. In this way, the adoption of the green technologies developed by group K is promoted. In the sense of "issue linkage", the coalition K may also introduce separate incentive mechanisms — as additional financial aid as a carrot or even trade restrictions as envisaged, e.g., by Nordhaus [2015] a stick — to ensure broad dissemination of its green innovation.

Moreover, the countries outside the coalition K may notice that their unwillingness to apply the new technology can undermine the willingness of coalition K to make the R&D-efforts and to form a technological coalition. To prevent this undesirable outcome, the outsiders also may form a separate coalition in which they commit themselves to adopt the improved technology.

7. An Example

We now specifically assume that $w = 1$ and that all countries have the Cobb-Douglas utility function $u(x_i, G) = x_i^\rho G$. For some marginal rate of substitution α, the expansion path is given by $e(G, \alpha) = \frac{\rho}{\alpha} G$ which gives $e_1(G, \alpha) = \frac{\rho}{\alpha}$ and $e_2(G, \alpha) = -\frac{\rho}{\alpha^2} G$. According to equation (4), the symmetric Nash equilibrium at $a_0 = 1$ is given by the public good supply level $\hat{G}(1) = \frac{n}{n\rho+1}$, the private good consumption levels $\hat{x}(1) = \frac{n\rho}{n\rho+1}$ and country-specific public good contributions $\hat{z}(1) = \frac{1}{n\rho+1}$. Since $\gamma_1 = \rho$ and $\gamma_2 = -\frac{n\rho}{n\rho+1}$ we get

$$\hat{G}' = \frac{k + m\beta - k\kappa_k}{n\rho + 1} \tag{18}$$

$$\hat{u}'_K = \frac{(\rho + 1)(k + m\beta - k\kappa_k) - n\rho}{n\rho + 1} \tag{19}$$

$$\hat{u}'_L = \frac{(\rho + 1)(k + m\beta - k\kappa_k)}{n\rho + 1} \tag{20}$$

$$\hat{u}'_M = \frac{(\rho + 1)(k + m\beta - k\kappa_k) - n\rho\beta}{n\rho + 1}. \tag{21}$$

We now especially look at equation (19) and consider the extreme case when there is only a single pioneering country, i.e., $k = 1$. The innovation is profitable for this country if its R&D-costs are below a certain threshold level, i.e.,

$$\kappa_1 < 1 + m\beta - \frac{n\rho}{\rho + 1}. \tag{22}$$

Now let either $m = 0$ or $\beta = 0$ so that there are no technological spillovers. Then, even if the innovation is completely costless i.e., $\kappa_1 = 0$ the potentially pioneering country has no incentive to increase its public good productivity if $1 - \frac{n\rho}{\rho+1} < 0$ or, equivalently, $n > \frac{\rho+1}{\rho}$, which is always the case if the total number of countries is sufficiently large. A single country then would even have an incentive to choose a technology with higher abatement costs, which is the paradoxical effect described by Buchholz and Konrad [1994]. By this strategy, this country would motivate the outsiders to increase their public good contributions on which it could then take the benefit as a free-rider.

But if in contrast there is a technological spillover the innovation will be profitable for the country if the technological spillover extends to sufficiently many countries, i.e., if

$$m > \frac{(n-1)\rho - 1}{(\rho + 1)\beta} \tag{23}$$

Some values for m which satisfy condition (23) exist if the right-hand side of this inequality is smaller than $n - 1$, i.e., if the spillover is sufficiently strong so that

$$\beta > \underline{\beta} := \frac{1}{n-1}\left(\frac{n\rho}{\rho + 1} - 1\right) \tag{24}$$

holds. In the case of a perfect spillover i.e., if $\beta = 1$ condition (24) is always fulfilled since $\underline{\beta} < 1$. The example thus clearly illustrates how a single country's incentive to innovate depends on the number of followers and the strength of the spillover effect. For a further specification, consider the case where $n = 10$ and $\rho = 1$. Then without a spillover, a costless marginal increase of public good productivity does not pay for a single country. But if there is a spillover with $\beta = 1$, which benefits at least five other countries, condition (24) implies that the innovation becomes worthwhile for the country which undertakes it. Moreover, the lower threshold for the productivity parameter which follows from condition (24) is $\underline{\beta} = \frac{4}{9}$.

Concerning the issue of internal stability of a coalition, as in Section 5 we consider the case with $l = 0$. According to (19), the size k^* of an internally stable coalition if existent is given by the smallest integer k for which

$$(1-\beta)k + n\left(\beta - \frac{\rho}{\rho+1}\right) > k\kappa_k \tag{25}$$

holds. It is well possible that $k^* = n$, i.e., that a grand technological coalition is stable. To make this precise let, first of all, $\beta = k\kappa_k = 0$. Then condition (25) reduces to

$$k > \frac{n\rho}{\rho+1}. \tag{26}$$

For a value of $\rho = 19$ and $n = 10$, condition (26) is clearly fulfilled for $k = n = 10$, but not for any $k < 9$. By a continuity argument, this is also true if β and $k\kappa_k$ are sufficiently small. A further analysis of the conditions for coalition stability in our framework and, in particular, the impact of the technology spillover [see El-Sayed and Rubio, 2014] will be a topic of future research.

8. Conclusion

In this chapter, we have shown how the provision of a global public good such as climate protection may be improved through unilateral action of a group of countries which collectively carries out a green technological innovation lowering the costs of providing the global public good, i.e., in the case of climate change the costs of greenhouse gas abatement. The success of such a specific form of leading behavior is not only more likely if the cooperating coalition is large but also if there is a steep rise of public good productivity in as many other countries as possible, i.e., if the technological spillover effect is strong both at the intensive and at the extensive margin. The basic message of this chapter is that it is not only favourable for the climate but also for the coalition if these follower countries get free access to the improved technology. They thus receive some indirect donation from the coalition so that such technological transfers serve as a substitute for climate finance proper. To put it differently: The knowledge, which is generated by R&D-activities, is a club good whose use can be restricted through laws for the protection of intellectual property. From the perspective of the developed countries it may, however, pay to transform this club good into a true public good with open access for all.

However, these recipient countries benefit less from the innovation than the complete outsiders that stick to the old high-cost technology. This creates an incentive problem for technology adoption so that it may become necessary to complement the unilateral R&D-policy by additional mechanisms to ensure a far-reaching diffusion of newly developed green technologies, i.e., to enlarge this special "climate club".[8] A discussion of appropriate strategies lies outside the scope of this theoretical work but should deserve a separate more empirically oriented treatment.

At the theoretical level the analysis presented in this paper could be extended and refined in different ways, especially by dropping the assumption that all countries are identical. So one could ask how the results are changed when it is taken into account that countries are neither equally rich nor have the same preferences for the public good. Moreover, a scenario might be explored in which the coalition K does not only decide on their R&D-efforts but also about whether it should also cooperatively determine its public contribution taking the reactions of the outsiders into account. The treatment of these issues will be reserved to future research.

Acknowledgments: We thank Anil Markandya and Todd Sandler for their comments and suggestions, which helped to improve our paper. Comments and suggestions from participants of presentations at the University of Hohenheim, the 22nd Annual EARE Conference in Zurich 2016 and the Workshop in Memory of Richard Cornes at the ANU Canberra also helped improve the paper. We also gratefully acknowledge financial support by the German Federal Ministry of Education and Research BMBF through the research project

[8]Referring to a different scenario of global climate policy, this notion has been coined by Nordhaus [2015].

ECCUITY (FKZ 01LA1104B) as part of the integrated programme "Ökonomie des Klimawandels" ("Economics of Climate Change").

References

Andreoni, J. 1988. Privately Provided Public Goods in a Large Economy: The Limits of Altruism. *Journal of Public Economics* 35, 57–73.

Barrett, S. 2003. *Environment and Statecraft: The Strategy of Environmental Treaty-Making.* Oxford, UK: Oxford University Press.

Barrett, S. 2006. Climate Treaties and "Breakthrough" Technologies. *American Economic Review.* Papers and Proceedings 96, 22–25.

Bergstrom, T., L. Blume and H. Varian. 1986. On the Private Provision of Public Goods, *Journal of Public Economics* 29, 25–49.

Buchholz, W., C. Haslbeck and T. Sandler. 1998. When Does Partial Cooperation Pay? *FinanzArchiv* 55, 1–20.

Buchholz, W. and K. A. Konrad. 1994. Global Environmental Problems and the Strategic Choice of Technology. *Journal of Economics* 60, 299–321.

Buchholz, W. and K. A. Konrad. 1995. Strategic Transfers and Private Provision of Public Goods. *Journal of Public Economics* 57, 489–505.

Buchholz, W. and W. Peters. 2001. The Overprovision Anomaly of Private Public Good Supply. *Journal of Economics* 74, 63–78.

Buchholz, W. and T. Sandler. 2017. Successful Leadership: Incorporating Behavioural Approaches into the Theory of Public Good Provision to appear in *Environmental and Resource Economics.*

Buchholz, W., R. Cornes, W. Peters and D. Rübbelke. 2015. Pareto Improvement through Unilateral Matching of Public Good Contributions: The Role of Commitment. *Economics Letters* 132, 9–12.

Buchner, B., A. Falconer, M. Hervé-Mignucci, C. Trabacchi and M. Brinkmann. 2011. *The Landscape of Climate Finance.* CPI Report. Venice, Italy.

Carraro, C., J. Eyckmans and M. Finus. 2006. Optimal Transfers and Participation Decisions in International Environmental Agreements. *Review of International Organization* 1, 379–396.

Carraro, C. and D. Siniscalco. 1993. Strategies for the International Protection of the Environment. *Journal of Public Economics* 52, 309–328.

Cornes, R. 2016. Aggregative Environmental Games. *Environmental and Resource Economics* 63, 339–365.

Cornes, R. and R. Hartley. 2007. Aggregative Public Good Games. *Journal of Public Economic Theory* 9, 201–219.

Cornes, R. and T. Sandler. 1996, *The Theory of Externalities, Public Goods, and Club Goods* (2nd ed.). Cambridge, UK: Cambridge University Press.

Cornes, R. and T. Sandler. 2000. Pareto-Improving Redistribution and Pure Public Goods. *German Economic Review* 1, 269–286.

El-Sayed, A. and S. J. Rubio. 2014. Sharing R&D Investments in Cleaner Technologies to Mitigate Climate Change. *Resource and Energy Economics* 38, 168–180.

Finus, M. 2001. *Game Theory and International Environmental Cooperation.* Cheltenham, UK and Northampton, MA: Edward Elgar.

Finus, M. and D. Rübbelke. 2013. Public Good Provision and Ancillary Benefits: The Case of Climate Agreements. *Environmental and Resource Economics* 56, 211–226.

Golombek, R. and M. Hoel. 2005. Climate Policy under Technology Spillovers. *Environmental and Resource Economics* 31, 201–227.

Greenwald, B. and J. Stiglitz. 2014. *Creating a Learning Society. A New Approach to Growth, Development, and Social Progress.* New York, NY: Columbia University Press.

Harstad, B., F. Lancia and A. Russo. 2015. Compliance Technology and Self-Enforcing Agreements. CESifo Working Paper No. 5562.

Helm, C. and R. C. Schmidt. 2015. Climate Cooperation with Technology Investments and Border Carbon Adjustment. *European Economic Review* 75, 112–130.

Hoel, M. 1991. Global Environmental Problems: The Effects of Unilateral Action Taken by One Country. *Journal of Environmental Economics and Management* 20, 55–70.

Hoel, M. and A. de Zeuw. 2010. Can a Focus on Breakthrough Technologies Improve the Performance of International Environmental Agreements? *Environmental and Resource Economics* 47, 395–406.

Ihori, T. 1996. International Public Goods and Contribution Productivity Differentials. *Journal of Public Economics* 61, 139–154.

Newell, R. 2010. *International Climate Technology Strategies.* In Post-Kyoto International Climate Policy — Implementing Architectures for Agreement, J. E. Aldy and R. N. Stavins (eds.), Cambridge, UK and New York, NY: Cambridge University Press, pp. 403–438.

Nordhaus, W. 2015. Climate Clubs: Overcoming Free-Riding in International Climate Policy. *American Economic Review* 105, 1339–1370.

Popp, D. 2011. International Technology Transfer, Climate Change, and the Clean Development Mechanism. *Review of Environmental Economics and Policy* 5, 131–152.

Popp, D. 2014. *Climate-Friendly Technological Change for Developing Countries.* In The Oxford Handbook on the Macroeconomics of Global Warming, W. Semler and L. Bernard (eds.), Oxford, UK: Oxford University Press.

Rübbelke, D. T. G. 2011. International Support of Climate Change Policies in Developing Countries: Strategic, Moral and Fairness Aspects. *Ecological Economics* 70, 1470–1480.

Sandler, T. 2004. *Global Collective Action.* Cambridge, UK: Cambridge University Press.

Sinn, H. W. 2012. *The Green Paradox — A Supply-Side Approach to Global Warming.* Cambridge, MA and London, UK: The MIT Press.

Stavins, R. and Z. Ji (eds.). 2014. *International Cooperation: Agreements and Instruments.* In Climate Change 2014 — Mitigation of Climate Change. Contribution of Working Group III to the Fifth Assessment Report of the Intergovernmental Panel on Climate Change. Cambridge, UK and New York, NY: Cambridge University Press, pp. 1001–1082.

Stern, N. 2015. *Why Are We Waiting*? The *Logic, Urgency, and Promise of Tackling Climate Change.* Cambridge, MA and London, UK: The MIT Press.

Tirole, J. 2012. Some Political Economy of Global Warming. *Economics of Energy* & Environmental Policy 1, 121–132.

Urpelainen, J. 2013. Can Strategic Technology Development Improve Climate Cooperation? A Game-Theoretic Analysis. *Mitigation and Adaptation Strategies for Global Change* 18, 785–800.

Warr, P. G. 1982. The Private Provision of a Public Good is Independent of the Distribution of Income. *Economics Letters* 13, 207–211.

Part 3

Appraising Investments

Chapter 13

Mean-Reverting Lévy Jump Dynamics in the European Power Sector

Julien Chevallier[*,†] and Stéphane Goutte[‡,§]

[*] *Université Paris 8 (LED), 2 rue de la Liberté, 93526 Saint-Denis Cedex, France*
Email: julien.chevallier04@univ-paris8.fr
[†] *IPAG Business School (IPAG Lab), 184 Boulevard Saint-Germain, 75006 Paris, France*
[‡] *Université Paris 8 (LED), 2 rue de la Liberté, 93526 Saint-Denis Cedex, France*
Email: stephane.goutte@univ-paris8.fr
[§] *Paris School of Business, 59 Rue nationale, 75013 Paris, France*

Abstract

This chapter analyzes the interactions between the electricity and CO_2 (carbon) markets. In particular, we describe the dynamics of the fuel-switching price (from coal to gas) when taking into account carbon costs. Several stochastic processes are considered to model the fuel-switching price: (i) the Brownian motion, and (ii) the Lévy jump process. Besides, the probability density function is evaluated by considering the Gaussian case versus the Normal Inverse Gaussian (NIG) and the Variance Gamma (VG) distributions. The results unambiguously point out the need to resort to jump modeling techniques to model satisfactorily the fuel-switching price with evidence of heavy tails. The assumption of a Gaussian distribution is also clearly rejected in favor of its main competitors, whereas it is found that the NIG beats the VG distribution. Taken together, these empirical results convey implications for risk managers looking to forecast and hedge their utilities' production.

1. Introduction

With the Paris 21st Conference of the Parties (COP21) Agreement, climate finance has risen to new heights of strategic interest for businesses, consultancies, governments and international organizations alike. Climate finance will become a cornerstone of climate policy financing, that will become binding again in 2020, a few years after the Kyoto Protocol expired. Against this largely sucessful background of diplomatic stances, markets will reflect new realities and price interactions will be scrutinized more than ever by market practitioners and academics.

This chapter examines the interactions between the power and carbon markets. Indeed, the development of the carbon market since 2005 has impacted generators' dispatch decisions. Stylized facts on the power system reveal that (i) electricity generation from coal is cheaper than from natural gas, and (ii) coal is nearly twice as polluting as gas. Hence, the creation of a carbon price proportional to CO_2 emissions may lead to changes in the merit order of power plants, and make gas more profitable than coal. Given the relative importance of the power sector in the design of the European Union Emissions Trading Scheme (EU ETS),[1] the fuel-switching behavior of power companies may therefore translate into actual CO_2 emissions reduction (as intended by the regulator).

The central contribution of this chapter is to provide a jump-robust stochastic model of the fuel-switching mechanism in presence

[1]The EU Emissions Trading Scheme (EU ETS) was established in 2003 by the Directive 2003/87/EC, and launched for a trial period from 2005 to 2007 as a cost-effective scheme to comply with the EU's commitment to reducing its emissions of greenhouse gases by 8% below 1990 levels over the period 2008–2012 [Alberola *et al.*, 2008]. Across its 27 Member States, the EU ETS covers large plants from CO_2-intensive emitting industrial sectors: power generation, mineral oil refineries, coke ovens, iron and steel and factories producing cement, glass, lime, brick, ceramics, pulp and paper, and all combustion activities with a rated thermal input exceeding above 20 MW. Overall, this scheme covers more than 11,000 installations [Alberola *et al.*, 2009a].

of carbon costs. We augment the model by Çetin and Verschuere [2009] through the introduction of jumps in the underlying stochastic process of the fuel-switching behavior. The superiority of models including jumps[2] in the stochastic processes applied to the price formation of European Union Allowances (EUAs) has been recently documented by Chevallier and Sévi [2013].

Reasons for stochastic dynamics include expected demand for allowances and fluctuating fuel prices, which potentially lead to a change in the merit order of power production, fuel-switch, costs, weather changes, economic growth, etc. [Seifert *et al.*, 2008]. The Brownian motion assumption is usually intended to capture these uncertainties in a simplified way. Other popular stochastic processes include mean-reversion (Vasicek, Ornstein-Uhlenbeck) and/or jumps (Poisson, Lévy, Bernoulli).[3]

Identifying jumps in a stochastic process is important because if it has implications for risk management, option pricing, portfolio selection and has consequences for optimal hedging strategies. Indeed, when computed using simulation techniques, the quantiles sensibly differ when draws are from a continuous, or continuous plus jump distribution. Similarly, portfolio selection can be dramatically modified when some assets in the investment universe are potentially jumping [Cvitanić *et al.*, 2008; Liu *et al.*, 2003].

In recent years, many pure jump or jump-diffusion models have been suggested in the economic and statistical literatures to deal with

---

[2] A few papers have documented the stochastic properties of allowance prices, either to model the dynamic price equilibrium [Carmona *et al.*, 2009, 2010], or to address the question of derivatives valuation in emissions markets [Borovkov *et al.*, [2011]; Carmona and Hinz, 2011; Çetin and Verschuere 2009; Chesney and Taschini 2012; Hinz and Novikov, 2010].

[3] The rationale for mean-reversion is that the price of a commodity tends to be pulled back to its production cost plus a margin [Schwartz, 1997; Schwartz and Smith, 2000]. For commodities such as energy, the mean is determined by the marginal cost of production and the extent of demand. In the short-run, there can be deviations from this arithmetic mean, but in the long-run the price converges towards the marginal costs of production as a result of competition among the producers. That is why the modeling of energy prices using the mean-reversion process is quite common.

(possibly large) discontinuities in price processes; see the reference textbook by Cont and Tankov [2004]. Adding a jump component to a continuous component (leading to a mixture model) or considering only a jump component allows better data fit than with only a continuous component.[4] In addition, given the presence of jumps in the data, pure jump models are preferred by users because they are easier to handle for practical applications such as derivatives pricing or real-life problems such as the valuation of insurance contracts (see Ballotta [2005] or Kassberg *et al.* [2008]) or real-option valuation (Martzoukos and Trigeorgis, 2002).[5]

In this context, Martzoukos and Trigeorgis [2002] provide methodologies in the case of real options to deal with the issue of rare events that may be represented by jumps in the stochastic process.[6] Their results show the very significant impact of jumps, in particular for complex real options such as growth options or extension options. European utilities may also be interested in optimal hedging in the carbon market, where the hedging decision depends on the stochastic properties of the underlying security.

Our results extend previous empirical studies on fuel-switching. First, McGuiness and Ellerman [2008] have employed an econometric model of fuel-switching: When EUA prices are increasing, their study shows that the use of CCGT is encouraged, and the use of coal is lower.[7] Second, Delarue *et al.* [2009] have used a simulation model in a 10-zonal setting, representing the main part of the EU. They have demonstrated that fuel-switching occurred in the UK during the summer 2005. Third, based on a "switching band" analysis, Lujan *et al.*

---

[4]The large empirical evidence about financial data is discussed in Cont and Tankov [2004], among others.

[5]Other examples of pure jump or jump-diffusion models are provided in Jing *et al.* [2012]. The authors emphasize that such models have applications far beyond the financial domain.

[6]Distributions from several processes of different natures are plotted in Jing *et al.* [2012]. The distributional aspect of pure-jump or diffusion is strikingly different, thereby motivating investigation of the fine nature of the underlying process for financial as well as for non-financial (real-world) applications.

[7]The decrease in coal is met partly with gas.

[2011] have also highlighted the ability of power generators' to deliver CO_2 emissions abatement when the appropriate economic incentives are met (i.e., with a strong carbon price signal and relatively cheaper gas compared to coal).

It is found that jump models significantly outperform non-jump models. According to our numerical simulations, the Normal Inverse Gaussian (NIG) and Variance Gamma (VG) probability density functions beat the normal distribution. Hence, our results indicate that the best goodness-of-fit for the fuel-switching price is achieved by the Lévy jump model with a non-Gaussian distribution. Thereby, the findings contradict the previous study by Carmona *et al.* [2009] based on the Brownian motion and assumption of Gaussian distribution for the modeling of the fuel-switching price. When challenging the NIG versus the VG distributions, our empirical tests reveal that the most satisfactory distribution in terms of goodness-of-fit is represented by the NIG. There is also overwhelming evidence of heavy tails in the distribution of the fuel-switching price, hence we demonstrate the superiority of the methodologies used in our chapter. The implications cover a wide range of interests for power producers' dispatch decisions.

The remainder of the chapter is organized as follows. Section 2 provides background information. Section 3 describes the fuel-switching mechanism. Section 4 develops the stochastic model. Section 5 contains numerical simulations. Section 6 concludes.

2. Background

This section provides background information on carbon and electricity markets.

The central objective of the EU ETS is to provide incentives to energy-intensive industries (and among them to power generating companies) to internalize the costs of CO_2 emissions, by adopting low-carbon and more energy-efficient technologies, thereby leading to actual CO_2 emissions reduction. Each country is required to develop a National Allocation Plan (NAP), which, among other design features, addresses the national CO_2 emissions target. During Phase III

(2013–2020), the EU ETS has encountered a novelty in the allocation process with the introduction of auctioning.

One EUA allows emitting one metric ton of CO_2-equivalent in the atmosphere. All companies must meet their compliance requirement, or pay stiff penalties (€100/metric ton of CO_2 during Phase II). According to the World Bank (2012), EUA transactions in 2011 have reached €105.7 billion, representing an 11% increase year-on-year compared to 2010. A total of 7.9 billion EUAs were traded in the market in 2011, at an average price of €13.5/metric ton of CO_2. These figures demonstrate that the EU carbon market has grown to be a liquid commodity futures market.

As for any other market, the carbon market is driven by the equilibrium between supply and demand. The supply side is dependent on (i) regulatory decisions amending the design of the scheme [Chevallier *et al.*, 2009], (ii) the ability to bank and borrow [Alberola and Chevallier, 2009], (iii) the emissions-to-cap level [Ellerman and Buchner, 2008], (iv) macroeconomic variables [Alberola *et al.*, 2009a,b; Chevallier, 2009], and (v) the supply of Kyoto Protocol's projects mechanisms credits. The demand side, at least concerning the power sector, is highly volatile and depends on many fundamental factors such as energy prices and weather parameters [Alberola *et al.*, 2008; Arouri *et al.*, 2012].

The possibility to switch at low cost from one technology to another less polluting one (such as coal to gas) also appears central to the analysis of the power sector, which brings us to the importance of spreads between fuel prices. The ability of power generators to switch between their fuel inputs is expected to be the primary source of CO_2 emissions reduction in the power sector [McGuinness and Ellerman, 2008]. Indeed, when the carbon price is above the switching point, gas-fired power plants become more profitable than coal-fired ones.

The most constrained sector is the combustion sector, which represents the largest share of installations. The combustion sector contains many sub-activities: large electricity generation plants, district heating facilities (cogeneration when details were available) and other installations (based on the EU Classification NACE Rev.1

C-F). Electricity generation was constrained by -9% on average, e.g., as judging by the ratio of allocation in proportion of its emissions. The main reasons are the perceived ability of this sector to reduce CO_2 emissions at low abatement costs through fuel-switching opportunities between relative coal and gas prices.[8] Thus, there exists a significant potential for CO_2 emissions abatement in this sector, as documented by previous literature Delarue and D'haeseleer [2007, 2008] and Delarue *et al.* [2009].

Power companies are the most active participants in the market, and their behavior influences greatly the carbon price levels, in accordance with their energy mix. For arbitrage purposes, utilities look at their margin, notably when selling power forward. To do so, they closely follow the evolution of European energy prices in order to take advantage of improving margins. Hence, utilities are directly concerned with the fuel-switching mechanism. For instance, utilities watch out the economic incentives to produce electricity via gas plants (instead of coal), if the prices evolve in favor of gas use. In that case, they would emit less CO_2, and their demand for carbon allowances should decrease, pushing prices down. Besides, previous literature has highlighted the strong correlations between electricity and carbon prices [Bunn and Fezzi, 2009; Chen *et al.*, 2008].

In the next section, we detail the concept of fuel-switching in the electricity sector, as well as its use when introducing carbon costs.

3. Fuel-Switching in the Power Sector

In this section, we detail first the basic principle behind fuel-switching,[9] second its main driving factors, and third its economic content.

---

[8]Besides, the power sector appears less exposed to international (non-EU) competition in opposition to other industrial sectors such as iron and steel or cement production.

[9]Note that we aim at capturing the short-term effect of introducing emissions trading on power producers' fuel-switching behavior with emission factors unchanged and the plant efficiencies fixed. The analysis of fuel substitution with additional costs associated with longer-term investment decisions is left for further research.

3.1. *Basic principle*

In the power sector, electricity prices are set by the marginal generation technology. The different generation units are ranked by marginal costs from the cheapest technology to the most expensive one. This ranking is called "merit order", and depends on several parameters such as fuel prices, plant efficiencies, carbon intensity and carbon costs [Sijm *et al.*, 2005]. The introduction of carbon costs through climate policy instruments may therefore lead to changes in the merit order of power plants.

Without carbon costs, the marginal cost of electricity is computed as the ratio between fuel costs and plants' efficiencies:

$$MC = \frac{FC}{\eta} \tag{1}$$

with MC the marginal cost, FC fuel costs, and η the plant efficiency.[10]

The introduction of carbon costs modifies the marginal cost for each plant by introducing the emissions factor, which depends on the fuel and the amount of fuel burnt:

$$MC = \frac{FC}{\eta} + \frac{EF}{\eta}EC \tag{2}$$

with EF the emissions factor, and EC emissions costs. The switching point between a given coal plant and a given gas plant can be defined as the emissions cost that equalizes marginal costs, i.e., $MC_{gas} = MC_{coal}$. It represents the allowance cost that leads to switch between two plants in the merit order. This price depends on each plant's fuel costs, efficiency and emissions factor:

$$EC_{switch} = \frac{\eta_{coal}FC_{gas} - \eta_{gas}FC_{coal}}{\eta_{gas}EF_{coal} - \eta_{coal}EF_{gas}}. \tag{3}$$

If the EUA price is lower than this cost, generating electricity from coal is more profitable than from gas. Note that the allowance

[10]For the sake of simplicity, and to make the model amenable to empirical work, this chapter does not take into account costs associated with operations and maintenance, manpower, regulatory changes, etc.

cost of the switching point varies linearly with the coal-to-gas price ratio. If we note $r = \frac{FC_{gas}}{FC_{coal}}$ the fuel price ratio, then the switching point can be written as:

$$EC_{switch} = \frac{(\eta_{coal} r - \eta_{gas}) FC_{coal}}{\eta_{gas} EF_{coal} - \eta_{coal} EF_{gas}} = ar + b \tag{4}$$

with $a = \frac{\eta_{coal} FC_{coal}}{\eta_{gas} EF_{coal} - \eta_{coal} EF_{gas}}$ and $b = \frac{\eta_{gas} FC_{coal}}{\eta_{gas} EF_{coal} - \eta_{coal} EF_{gas}}$. Equation (4) shows that the switching point depends linearly on the fuel price ratio, but also that switching can occur even with zero allowance costs. The denominator is positive if $\frac{\eta_{gas}}{\eta_{coal}} > \frac{EF_{gas}}{EF_{coal}}$. This relationship is verified in practice: the emissions factor for gas-fired power plants is nearly half of the one for coal-fired power plants. Additionally, the efficiency of gas-fired power plants (especially CCGT) is usually higher than the one for coal-fired power plants.

Solving for $EC_{switch} = 0$ leads to $r = \frac{\eta_{gas}}{\eta_{coal}}$. If $r > \frac{\eta_{gas}}{\eta_{coal}}$, then $EC_{switch} > 0$, which means that the carbon cost for switching is positive. In the opposite case, fuel-switching could occur even in presence of zero carbon costs. When the coal-to-gas price ratio is smaller (or equal to) the coal-to-gas efficiencies ratio, fuel-switching may occur without carbon costs. Hence, for a given positive coal price, there exists a gas price which is low enough to influence power producers' fuel-switching behavior, even with a zero carbon price. Mathematically, this condition is met when the coal-to-gas ratio is equal to the ratio of the gas and coal efficiencies [Delarue *et al.*, 2009].

Knowing the various plant outputs, these computational steps allow to draw the emissions cost profile leading to the use of gas instead of coal depending on the load. At base-load, nuclear, coal and renewables are typically used to meet demand, not gas. The introduction of CO_2 costs may provide incentives for generators to use gas instead of coal at the switching point. Depending on the load and the number of gas units available, the emissions cost profile will differ.

In theory, the switching point could occur between all the technologies available for power generation. However, in practice, the

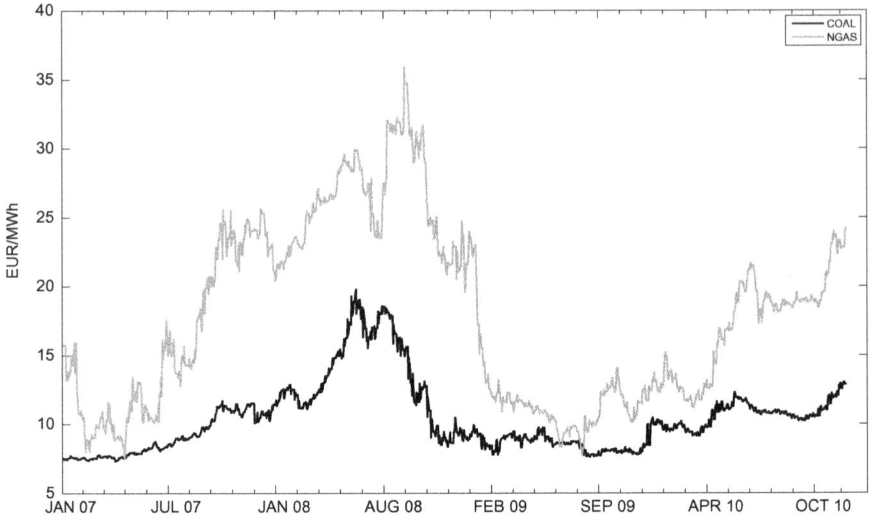

Figure 1. EEX Coal and Natural Gas Prices (in EUR/MWh) from 1 January 2007 to 31 December 2010.

main abatement opportunities are expected to come from the switching from coal to gas. Switching from coal to oil (or from oil to gas) is possible, but it appears very limited in the European electricity generation mix, and it is usually more expensive. Switching from coal to nuclear (or gas to nuclear) also appears unlikely, since nuclear energy is not flexible and has to operate at high-load to be profitable.[11]

Figure 1 displays two curves. First, the natural gas price used (expressed in EUR/MWh) is the futures Month Ahead price negotiated on EEX from January 1, 2007 to December 31, 2010. It represents one of the most liquid gas contracts in Europe, and has a major influence on the price that European consumers pay for gas. As such, the EEX Natural Gas futures price represents one of the best proxies of the European gas market price determined close to end-users. Second, the coal price is the EEX coal futures Month Ahead price API#2 (expressed in EUR/MWh), which is the major

[11]Therefore, nuclear energy is used to meet base-load demand (not peak load demand) as its power output cannot be modulated.

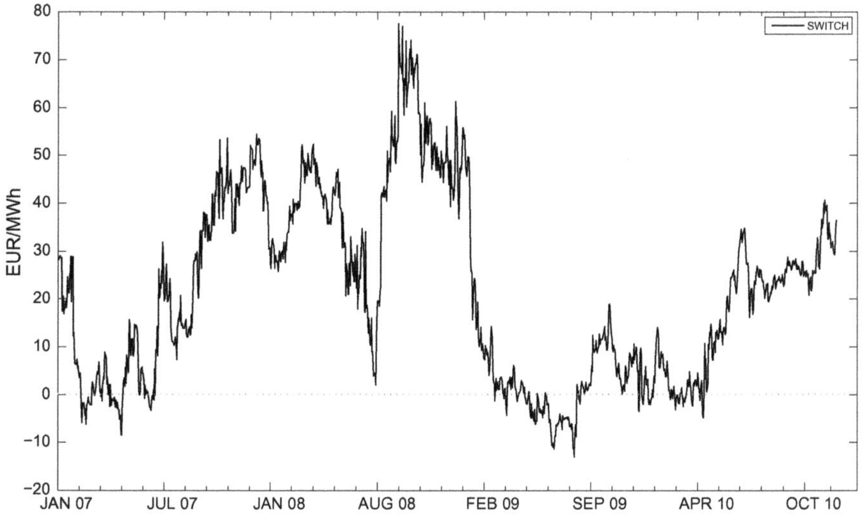

Figure 2. Switch Price (in EUR/MWh) from 1 January 2007 to 31 December 2010.

imported coal contract in northwest Europe. We notice immediately the close behavior of the two time series, with specific periods during which fuel-switching may have occurred in the EU ETS, especially during January–July 2007 and February–September 2009.

Following Equation (4), the computation of the switch price between EEX Coal and Natural Gas prices yields to the curve represented in Figure 2. As a by-product, the Switch Price is also expressed in Euro/MWh.

To obtain the switch price, we have plugged the values $cost_{ngas} = 0.5$ as the production cost of one MWh of electricity based on net CO_2 emissions of gas (e.g., the efficiency of the typical gas plant, expressed in EUR/MWh), $cost_{coal} = 0.4$ the production cost of one MWh of electricity based on net CO_2 emissions of coal (e.g., the efficiency of the typical coal plant, expressed in EUR/MWh), $tCO2_{coal} = 364.68$ the emissions factor (expressed in $kgCO_2eq/MWhp$) of a conventional coal-fired plant, and $tCO2_{ngas} = 210.96$ the emissions factor (expressed in $kgCO_2eq/MWhp$) of a conventional gas-fired plant (see Delarue and D'haeseleer [2007, 2008] and Delarue *et al.* [2008]).

As long as the carbon price lies below this switching price, coal plants are more profitable than gas plants — even after taking carbon costs into account. Against the dotted horizontal line, we confirm visually the occurrence of fuel-switching during brief sub-periods: January 2007–July 2007 and February 2009–April 2010.

Let us examine in more details the effects of EUA prices on power producers' fuel-switching behavior in the next section.

3.2. *Factors influencing fuel-switching*

Several factors may impact fuel-switching opportunities. Previous literature has identified the influence of the load, CO_2 prices and fuel prices as being the main drivers behind the fuel-switching potential.[12]

The first factor influencing fuel-switching is represented by fuel prices. The switching point may be seen as the EUA price at which unused available gas-fired capacity is substituted for coal-fired generation. For any given couple of coal- and gas-fired plants connected to the electricity grid, there is a single switching point. However, there are many pairs of plants with different efficiencies and fuel prices, which leads to a "switching band". The lower bound of the band corresponds to the substitution of the most efficient and lowest cost unused gas-fired plant with the least efficient and highest cost coal-fired plant in service. The upper bound is the substitution of the most efficient and lowest cost coal by the least efficient highest cost gas [Delarue *et al.*, 2009]. A low coal-to-gas ratio encourages fuel-switching because the switching band is low.

Second, the fuel-switching potential is very dependent on the load. At full load, during winter peak hours for instance, all the plants are running, so that no opportunity to switch from coal to gas exists. Switching occurs only if coal-fired power plants are running, while some gas-fired power plants are available to replace them. These

[12]While fuel price relations can lead to lower levels of emissions than can a carbon price (with a different set of fuel prices), a carbon price can be expected to lead to lower emissions whatever the set of fuel prices. We wish to thank a referee for this remark.

conditions may be found when the load is relatively low and mostly met by coal-fired plants, i.e., during weekends, nights and summers. With adequate economic incentives, available gas-fired units can be used instead of coal-fired units. Fuel prices are usually more expensive during the winter (especially natural gas). This seasonal variability constitutes one of the main reasons why fuel-switching is more likely to occur during the summer. Consequently, the fuel-switching potential varies throughout the year, depending on the season (winter or summer), the time of the week (weekday or weekend), and the period of the day (day or night).[13]

The EUA price constitutes the third fundamental factor affecting the potential for fuel-switching. High EUA prices are more likely to fall within the switching band. Delarue *et al.* [2008] compare the potential reduction at various levels of EUA prices in Europe. At €20/metric ton of CO_2, there exists a significant potential for emissions reduction during the summer. This result is in line with Sijm *et al.* [2005], who calculated that the breakeven price of CO_2 in 2005 was around €18.5/metric ton of CO_2. €60/metric ton of CO_2 allow a significant reduction throughout the year, while €120/metric ton of CO_2 lead to a constant reduction during the whole year. Around €150/metric ton of CO_2, all the switching opportunities are used, and no further abatement can be reached using fuel-switching only.

3.3. *Economic content*

The "fuel-switching price" has been defined in Equation (4) as the CO_2 price which makes a gas-fired power plant economically equally profitable as a coal-fired power plant. It exists for every pair of gas and coal plants. Consequently, the fuel-switching price conveys a meaningful content for economists.

[13]Previous literature documents that fuel-switching took place in Germany during the summer of 2005, when gas prices were cheap and the load sufficiently low [Delarue *et al.*, 2008; Ellerman and Feilhauer, 2008]. Other studies highlight the greater potential for fuel-switching at low cost in the summer. For instance, Delarue and D'haeseleer [2007] show that a 9.5% reduction is achievable in the Belgian-based electrical power system.

The impact of climate policy on power generation fuel choices has attracted a lot of attention among researchers and policy makers. A noted by Ellerman and Buchner [2008], an understanding of the fuel-switching behavior among fossil-fueled power plants is essential in assessing compliance issues within the EU ETS. First, it provides us with a clear representation of what actually occurs in the EU power system in terms of potential CO_2 emissions abatement. This is the key delivery variable by which the DG CLIMA of the EU Commission evaluates the relative "success" of its ETS [Convery and Redmond, 2007]. These estimates are also central to analysts in the competitive sector of power production, and to brokers as well as an early warning signal to capture anticipations of carbon price swings (as in the April 2006 episode, see Alberola *et al.* [2008]). Further on this — from a market practitioner's standpoint — the creation of the EU ETS induces mechanically important fuel-switching behaviors in the short-run, and an understanding of this behavior would assist in forecasting future price developments in the carbon allowance market.

Second, the EU 27 has a great potential for CO_2 emissions abatement in the power sector through fuel-switching [Sijm *et al.*, 2005]. CO_2 emissions reduction may vary significantly depending on the country, the carbon intensity of the power sector, and the ability to switch between different fuels. The switching from coal to gas is one of the lowest cost sources of abatement expected in power generation. It is relatively easy to implement, and leads to more efficient use of gas plants such as Combined Cycle Gas Turbine [CCGT, see Delarue and D'haeseleer, 2007]. Creti *et al.* [2012] proxy the abatement opportunities through fuel switching by calculating the price of allowances at which the marginal costs of gas- and coal-fired power plants are equal, coined "switching price". Assuming that each operator has the possibility to switch from high to low inputs, fuel price differentials affect the technology used to produce electricity, thus fuel demand and consequently EUA demand as well. In particular, if the EUA is below the switching price, there is an economic advantage to use coal-fired power plants even in carbon-constrained frameworks.

Third, encouraging fuel-switching through the creation of a carbon price appears as a cost effective and almost natural means to cut CO_2 emissions (Ellerman and Buchner [2007, 2008]). The carbon price provides incentives for generators to switch from coal to gas, simply because it minimizes the marginal cost of producing electricity. However, the carbon price alone may not be enough to promote renewables for instance, which are still more expensive than fossil fuel generation. Therefore, the carbon price has to be seen as a component of a CO_2 abatement opportunity in the EU comprehensive strategy to tackle climate change.[14]

Fourth, the interaction of fuel, carbon and electricity prices that is at stake through the fuel-switching mechanism can be used to evaluate the efficiency of the corresponding markets. If fuel and carbon prices were to exhibit a low explanatory power for the electricity price in a system dominated by thermal units, this would indicate that pricing is not based on marginal cost. With the advent of the EU ETS and the important role of (short-run) power generation fuel substitution in the scheme, this topic constitutes an active area of research in the energy economics literature, from which we draw selected works. Pettersson *et al.* [2012] analyze the role and the nature of price-induced switching behavior between fossil fuels (e.g., coal, oil, and natural gas) in the western European power sector. They find that the prospects for short-term fuel-switching are significant, and that the introduction of carbon pricing in the electric power sector may provide incentives to improve also the future flexibility of coal use. Based on a theoretical model, Falbo *et al.* [2013] consider the case of an electricity producer comparing different generation technologies, and analyze whether switching to less polluting technologies is a profitable investment alternative. Their empirical evidence concludes that high allowance prices may generate a shift in production from less to more polluting technologies.

[14]This strategy has to provide incentives for the development of renewable energies, such as renewable heat generation and Combined Heat and Power targets, clean technologies such as fuel cells, and promote energy efficiency.

Fifth, a stochastic fuel-switching model allows for various interesting applications for economists [Zachmann, 2013]. Since multi-commodity options are quite common in electricity markets (e.g., dark, spark and clean spreads), the ability of the stochastic model to statistically represent the co-movements of these commodities could prove helpful for evaluating corresponding options. Our objective lies in documenting the presence of jumps, and therefore the need to resort to Lévy jump processes. The gain in using a jump process instead of a continuous one (as in the present literature) is twofold: (i) we capture spikes in typical power plants running on gas or coal, and (ii) among jump processes, we document that the NIG process is again the best distribution in the energy sector. A natural use of our results lies in hedging derivatives products on power markets. Indeed, we can elaborate an optimal hedging strategy based on the jump modelization.

For these various reasons, the economist's task to accurately model the fuel-switching variable appears of utmost importance. Hence the strategic value of fitting the adequate stochastic process to the fuel-switching price, as examined in the next section.

4. The Stochastic Model

Let $(\omega, \mathcal{F}, P)$ be a filtered probability space, and T be a fixed terminal time horizon. We propose in this chapter to model the fuel-switching price by using continuous-time stochastic jump diffusions. Indeed, the central concern to obtain risk measures in energy derivatives lies in finding a model that fits adequately the data-generating process of the time series under consideration. The normal distribution is mostly used for the modeling of financial log-returns. However, we can readily observe that the fuel-switching price pictured in Figure 2 exhibits salient features departing from Gaussian assumption such as fat tails (or semi-heavy tails), excess skewness, and jumps.

That is why we propose in this chapter to model the fuel-switching price defined by (4) as a continuous time stochastic process denoted by X_t. It will follow a mean-reverting Lévy jump model.

4.1. *The stochastic model*

We begin by the definition of a Lévy process.

Definition 1. A Lévy process L_t is a stochastic process such that

1. $L_0 = 0$.
2. For all $s > 0$ and $t > 0$, we have that the property of stationary increments is satisfied. i.e., $L_{t+s} - L_t$ as the same distribution as L_s.
3. The property of independent increments is satisfied. i.e., for all $0 \leq t_0 < t_1 < \cdots < t_n$, we have that $L_{t_i} - L_{t_{i-1}}$ are independent for all $i = 1, \ldots, n$.
4. L has a Cadlag paths. This means that the sample paths of a Lévy process are right continuous and admit a left limits.

Remark 2. In a Lévy process, the discontinuities occur at random times.

Since our aim is to find the model that provides the best goodness-of-fit to the fuel-switching price historical values, we conduct a "horse-race" between the three competing models. Thus, the stochastic process X will be given by one of the solution of the following stochastic differential equation:

Continuous process

$$dX_t = \kappa \left(\theta - X_t\right) dt + \sigma dW_t \tag{5}$$

with parameters κ, θ in $\mathbb{R}$ and σ in $\mathbb{R}^+$ and where W_t is a Brownian motion.

NIG Lévy jump process

$$dX_t = \kappa(\theta - X_t)dt + \sigma dL_t \tag{6}$$

with parameters κ,θ in $\mathbb{R}$ and σ in $\mathbb{R}^+$ and where L_t follows a NIG process.

VG Lévy jump process

$$dX_t = \kappa\,(\theta - X_t)\,dt + \sigma dL_t \tag{7}$$

with parameters κ, θ in $\mathbb{R}$ and σ in $\mathbb{R}^+$ and where L_t follows a Variance Gamma (VG) process.

Remark 3. We have in our three models that

- κ denotes the mean-reverting rate;
- θ denotes the long-run mean;
- σ denotes the volatility of X.

Proposition 1. *For all $t \in [s, T]$, the solution of the stochastic differential equations* (5), (6) *and* (7) *are given by*

$$X_t = X_s e^{-\kappa(t-s)} + \theta(1 - e^{-\kappa(t-s)}) + \sigma \int_s^t e^{-\kappa(t-u)} dZ_u \tag{8}$$

where Z is respectively a Brownian motion, a NIG or a VG process for respectively the models (5), (6) *and* (7).

4.2. *Estimation*

In practice, we observe when the price in fixed times $0 = t_0 < t_1 < \cdots < t_n = T$, with $\Delta_t = t_{k+1} - t_k$ is a constant, our stochastic models give

$$X_{t_{k+1}} - X_{t_k} = m - aX_{t_k} + s\varepsilon_{t_k} \tag{9}$$

with

$$m = (1 - e^{-\kappa\Delta_t})\theta$$

$$a = 1 - e^{-\kappa\Delta_t}$$

$$s = \sqrt{\frac{1 - e^{-2\kappa\Delta_t}}{2\kappa}}\sigma$$

and ε_{t_k} are independent and identically distributed random variables with mean zero and standard deviation one. More precisely, we have for any $k = 0, \ldots, n-1$

$$\varepsilon_{t_k} = \frac{1}{s} \int_{t_k}^{t_{k+1}} \sigma e^{-\kappa(t_{k+1}-u)} dZ_u. \tag{10}$$

Moreover, we distinguish the models as follows:

- In the case of Model (5), the process Z is a standard Brownian motion W, thus the noise ε_{t_k} follows a $\mathcal{N}(0,1)$ distribution for all $k = 0, \ldots, n-1$.
- In the case of Models (6) and (7), where the process Z follows a NIG or VG distribution, the continuous time model is not discretized exactly. Thus, we make the approximation (valid for small values of $\kappa\Delta_t$) that the noise ε_{t_k} is independent and identically distributed according to the NIG or VG distributions followed by L_1. The random variable L_1 is supposed to be mean-zero with standard deviation one, so that ε_{t_k} is supposed to be centered and standard.

For simplicity of the notation, we will denote, in the sequel, the observation X_{t_k} by X_k.

Remark 4. The continuous time Brownian motion model (5) is clearly a reduced case of the more general Lévy jump model. Indeed, a Brownian motion is a particular Lévy process where the jump part in the Lévy decomposition is null.

We recall now some distribution result of the NIG and VG distributions. These two distributions are particular cases of the Hyperbolic generalized distribution (HG) introduced by Barndorff–Nielsen and Halgreen [1977]. The density function of an HG is given by

$$\begin{aligned} f_{HG}(x; \lambda, \alpha, \beta, \delta, \mu) &= a(\lambda, \alpha, \beta, \delta)(\delta^2 + (x-\mu)^2)^{(\lambda-1/2)/2} \\ &\quad \times K_{\lambda-1/2}(\alpha\sqrt{\delta^2 + (x-\mu)^2}) \exp(\beta(x-\mu)), \end{aligned} \tag{11}$$

where

$$a(\lambda, \alpha, \beta, \delta) = \frac{(\alpha^2 - \beta^2)^{\lambda/2}}{\sqrt{2\pi}\alpha^{\lambda-1/2}\delta^\lambda K_\lambda(\delta\sqrt{\alpha^2 - \beta^2})} \tag{12}$$

is a normalization constant and K_ν is the third Bessel kind function with index ν. It can be represented with the following integral

$$K_\nu(z) = \frac{1}{2}\int_0^\infty y^{\nu-1}\exp\left(-\frac{1}{2}z(y+y^{-1})\right)dy.$$

For a given real ν, the function K_ν satisfies the differential equation given by

$$x^2y'' + xy' - (x^2+\nu^2)y = 0.$$

Let $\gamma = \sqrt{\alpha^2-\beta^2}$, then the parameters of this density satisfy:

$$\begin{cases} \delta \geq 0, & \alpha > 0, \quad \gamma > 0 \quad \text{if } \lambda > 0, \\ \delta > 0, & \alpha > 0, \quad \gamma > 0 \quad \text{if } \lambda = 0, \\ \delta > 0, & \alpha \geq 0, \quad \gamma \geq 0 \quad \text{if } \lambda < 0. \end{cases} \tag{13}$$

We are now interested in two particular cases of this distribution. $\lambda = -1/2$. We have $\delta > 0$, $\alpha \geq 0$ and $\gamma \geq 0$.

4.2.1. *Normal inverse gaussian (NIG)*

Taking $\lambda = -1/2$ in (11), we have $\delta > 0$, $\alpha \geq 0$ and $\gamma \geq 0$ and the process L follows now a NIG distribution. The density function of a NIG variable $NIG(\alpha,\beta,\delta,\mu)$ is given by

$$f_{NIG}(x;\alpha,\beta,\delta,\mu) = \frac{\alpha}{\pi}\exp\left(\delta\sqrt{\alpha^2-\beta^2}+\beta(x-\mu)\right)\frac{K_1\left(\alpha\delta\sqrt{1+(x-\mu)^2/\delta^2}\right)}{\sqrt{1+(x-\mu)^2/\delta^2}} \tag{14}$$

where K_ν is always the third Bessel kind function with index ν. This class of distribution is stable by convolution as the classic normal distribution, i.e.,

$$NIG(\alpha,\beta,\delta_1,\mu_1) * NIG(\alpha,\beta,\delta_2,\mu_2) = NIG(\alpha,\beta,\delta_1+\delta_2,\mu_1+\mu_2).$$

Lemma 5. *If* $X \sim NIG(\alpha, \beta, \delta, \mu)$ *then for any* $a \in \mathbb{R}^+$ *and* $b \in \mathbb{R}$, *we have that*

$$Y = aX + b \sim \left(\frac{\alpha}{a}, \frac{\beta}{a}, a\delta, a\mu + b\right).$$

We have that the log cumulative function of a NIG variable is given by

$$\phi^{NIG}(z) = \mu z + \delta\left(\sqrt{\alpha^2 - \beta^2} - \sqrt{\alpha^2 - (\beta + z)^2}\right),$$
$$\text{for all } |\beta + z| < \alpha. \tag{15}$$

The first moments are given by

$$\mathbb{E}[X] = \mu + \frac{\delta\beta}{\gamma}, \qquad \text{Var}[X] = \frac{\delta\alpha^2}{\gamma^3}, \tag{16}$$

$$\text{Skew}[X] = \frac{3\beta}{\alpha(\gamma\delta)^{\frac{1}{2}}} \quad \text{and} \quad \text{Kurt}[X] = 3\frac{1 + \frac{4\beta^2}{\alpha^2}}{\delta\gamma}, \tag{17}$$

with $\gamma = \sqrt{\alpha^2 - \beta^2}$. And finally the Lévy measure of a NIG($\alpha, \beta, \delta, \mu$) law is

$$F_{NIG}(dx) = e^{\beta x}\frac{\delta\alpha}{\pi|x|}K_1(\alpha|x|)\,dx. \tag{18}$$

Remark 6. Each parameter in $NIG(\alpha, \beta, \delta, \mu)$ distributions can be interpreted as having a different effects on the shape of the distribution:

- α — tail heaviness of steepness.
- β — symmetry.
- δ — scale.
- μ — location.

4.2.2. *Variance gamma (VG)*

The VG density was initiated by Madan *et al.* [1998] and corresponds to the particular of an Hyperbolic generalized distribution with $\delta = 0$.

In this case, we have $\lambda > 0$ and $\alpha > |\beta|$. The density function of the law $\mathrm{VG}(\lambda, \alpha, \beta, \mu)$ is given by

$$f_{VG}(x; \lambda, \alpha, \beta, \mu) = \frac{\gamma^{2\lambda}}{\sqrt{\pi}\Gamma(\lambda)(2\alpha)^{\lambda-1/2}}|x-\mu|^{\lambda-1/2} \times K_{\lambda-1/2}(\alpha|x-\mu|)e^{\beta(x-\mu)}, \quad \text{for all } x \in \mathbb{R}, \tag{19}$$

where Γ is the Gamma Γ function. The Laplace transform is given by

$$L_{VG}(v) = \mathbb{E}[\exp(vX)] = e^{\mu v}\left(\frac{\gamma}{\gamma_v}\right)^{2\lambda}, \quad |\beta + v| < \alpha, \tag{20}$$

where $\gamma_v = \sqrt{\alpha^2 - (\beta+v)^2}$. Its characteristic function is given by

$$\Psi_{VG}(u) = \log \mathbb{E}\big[\exp(iuX)\big] = i\mu u + \lambda \log\left(\frac{\alpha^2 - \beta^2}{\alpha^2 - (\beta+iu)^2}\right) \quad \text{for all } u \in \mathbb{R}. \tag{21}$$

Thus we deduce that

$$\mathbb{E}[X] = \mu + \frac{2\beta\lambda}{\gamma^2}, \quad \mathrm{Var}[X] = \frac{2\lambda}{\gamma^2}\left(1 + 2\left(\frac{\beta}{\gamma}\right)^2\right).$$

As the NIG distribution, the VG distribution is stable by convolution as the classic normal distribution:

$$VG(\lambda_1, \alpha, \beta, \mu_1) * VG(\lambda_2, \alpha, \beta, \mu_2) = VG(\lambda_1 + \lambda_2, \alpha, \beta, \mu_1 + \mu_2).$$

And finally the Lévy measure of a $\mathrm{VG}(\lambda, \alpha, \beta, \mu)$ law is

$$F_{VG}(dx) = \lambda \frac{\exp(\beta x)}{|x|} \exp(-\alpha|x|)\, dx. \tag{22}$$

4.3. *Parameters estimation*

We aim at estimating the set of parameters Θ that contains the mean-reverting diffusion parameters m and a, the volatility of the diffusion s, as well as the set of parameters of the distribution laws of Z. Thus, for the NIG model case, we have the following set of parameters

$$\Theta^{NIG} := \{m, a, s, \alpha, \beta, \delta, \mu\},$$

and for the VG model case

$$\Theta^{VG} := \{m, a, s, \lambda, \alpha, \beta, \mu\}.$$

We use the procedure of estimation of Chevallier and Goutte [2015]. The estimation procedure unfolds in two steps. First, we estimate the subset of parameter $\{m, a, s\}$ using a least squares method. Second, we estimate the second subset of parameters corresponding to the law distribution of Z (i.e., $\{\alpha, \beta, \delta, \mu\}$ in the NIG case) using a maximum likelihood method. This two-step approach greatly simplifies the task of the econometrician, as it reduces the optimization problem.

5. Empirical Application

Let us now apply the estimation strategy to the historical fuel-switching price given by Figure 2.

Descriptive statistics are reported in Table 1. We remark that the fuel-switching price fluctuates around a mean of €21.33/MWh. The occurrence of negative values indicates that some fuel-switching between gas and coal has actually been triggered during our sample period. Regarding the distribution of the raw time series, the fuel-switching price is characterized by an excess skewness and a kurtosis

Table 1. Descriptive Statistics.

Statistics	Data
Mean	21.3375
Median	20.1400
Minimum	−12.9800
Maximum	77.4800
Std	19.5542
Skewness	0.4336
Kurtosis	2.2116

coefficient that is far from three (i.e., the threshold for the normal distribution).[15]

5.1. *Model estimates*

Once estimated, the set of parameters for the NIG and VG case are reproduced in Tables 2 and 3. Similarly, the estimated parameters of the continuous time process are given in Table 4.

In the following Table 5, we give the values of the obtained log likelihood values for a panel of distributions. Furthermore, we also calculate the Akaike Information Criterion (AIC) and the Bayesian

Table 2. Estimated Parameters of the NIG Case.

NIG	α	β	δ	μ
	0.1290	0.0101	1.2792	0.1527

Table 3. Estimated Parameters of the VG Case.

VG	λ	α	β	μ
	0.6226	0.3971	−0.0007	0.2960

Table 4. Estimated Parameters of the Continuous Time Process.

Process	κ	θ	σ
	2.9787	21.8044	67.9655

[15] We have investigated the seasonality component D of the fuel-switching price. However, we were unable to find a significant seasonality component in the data. Indeed, when we look at the Figure 2, the seasonality part is not really evident. Seasonality tests are not reproduced in the chapter, and can be accessed upon request to the authors.

Table 5. Descriptive Statistics.

Distributions	**Log likelihood**	**BIC**	**AIC**
Logistic	3498	7011	7000
Normal	3671	7357	7347
Generalized extreme value	3790	7602	7586
Extreme value	4206	8427	8416
Rayleigh	3736	7478	7473
Generalized Pareto	5163	10348	10332
NIG	3358	6746	6724
Variance Gamma	3357	6742	6721

Information Criterion (BIC) which are given by

$$AIC = -2\ln(L(\Theta))+2*k \quad \text{and} \quad BIC = -2*\ln(L(\Theta^{(n)}))+k\ln(n), \tag{23}$$

where $L(\Theta)$ is the log-likelihood value obtained with the estimated parameters Θ found by our estimation procedure, k is the degree of freedom of each model and n the number of observations. We recall that the preferred model is the one with the minimum AIC or BIC value. Hence, lower are the BIC and AIC criteria better the fit of the model is.

We see clearly that NIG and VG distributions give the best results and that their values are very close.

5.1.1. *Estimation of the distribution law of the residuals*

Next, we investigate the best fit for the residuals, given by the law of the stochastic process Z defined in (10). The plot of the obtained residuals is given in Figure 3.

Further inspection of residuals tests are in order. They are conducted in Section 5.2.

5.2. *Goodness-of-fit*

This section contains further illustrative results on the goodness-of-fit offered by our jump model with NIG distribution or VG distribution vs. the standard Brownian motion with Normal distribution.

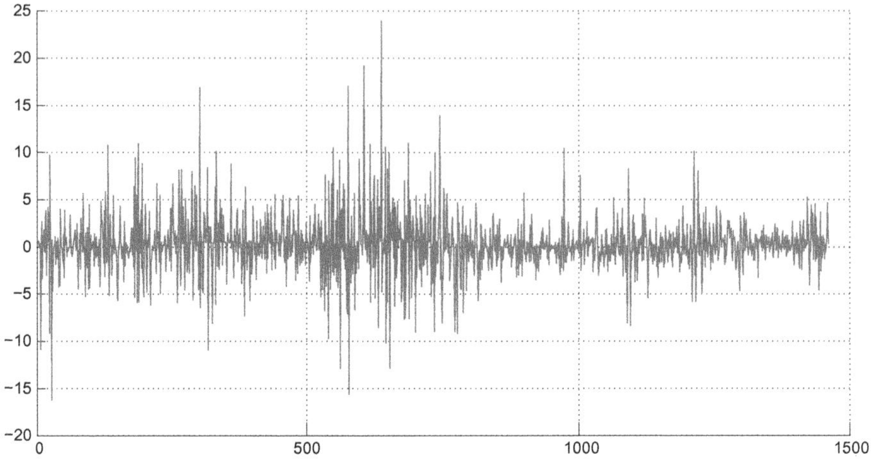

Figure 3. Residuals.

5.2.1. *Probability density function and QQ-plot*

Figure 4 displays the histogram of the historical residuals of the fuel-switching price, as modeled consistently with jumps in this chapter. The results contrast the probability density function in the standard "bell-shaped" Gaussian case (bottom), vs. the NIG case (top) or the VG case (middle). Fortunately, we assert that the residual's histogram exhibits a kurtosis coefficient that is significantly different from the normality assumption. On the contrary, we observe very clearly that the probability density function of the fuel-switching price residuals correspond more accurately to the NIG case or the VG case. Thus, the assumption that the model is driven by a Lévy jump process and not a Continuous Gaussian process is clearly visible in this figure.

Figure 5 contains additional empirical evidence to illustrate that the residuals do not follow a Normal distribution: in the bottom panel, the QQ-plot departs strongly from the Gaussian case. In the top panel, we can observe visually that the QQ-plot seems to follow the NIG distribution, and to fit much better the NIG distribution than the VG distribution. By contrasting the first two quadrants with the bottom one, we notice visually that the fuel-switching price is

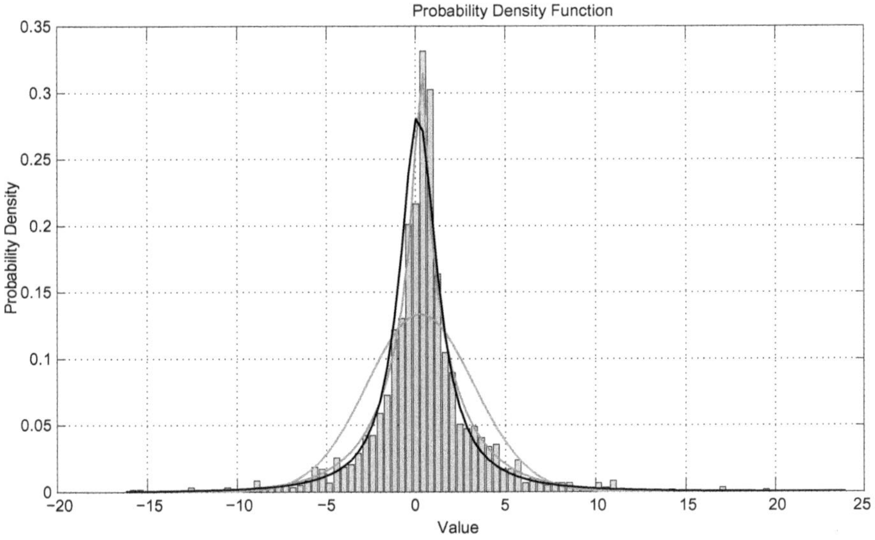

Figure 4. Histogram of Historical Residuals of Our Fuel-Switching Prices and the Corresponding NIG (top), VG (middle) and Normal (bottom) distributed probability density function.

characterized by heavy tails, strongly departing from the assumption of Gaussian assumption. This further motivates the need to resort to the methodologies used in our chapter.

5.2.2. *Kolmogorov-Smirnov test*

The Kolmogorov-Smirnov test is a form of minimum distance estimation, used as a parametric test of equality of one-dimensional probability distribution compared with a reference probability. The Kolmogorov-Smirnov statistic, D_n, where n is the sample size, quantifies a distance between the empirical distribution function of the sample $F_n(\Delta z)$ and the cumulative distribution function of a reference distribution $F(\Delta z)$. The empirical distribution function $F_n(\Delta z)$ for n i.i.d. observations Δz_i is defined as

$$F_n(\Delta z) = \sum_{i=1}^{n} 1(\Delta z_i \leq \Delta z),$$

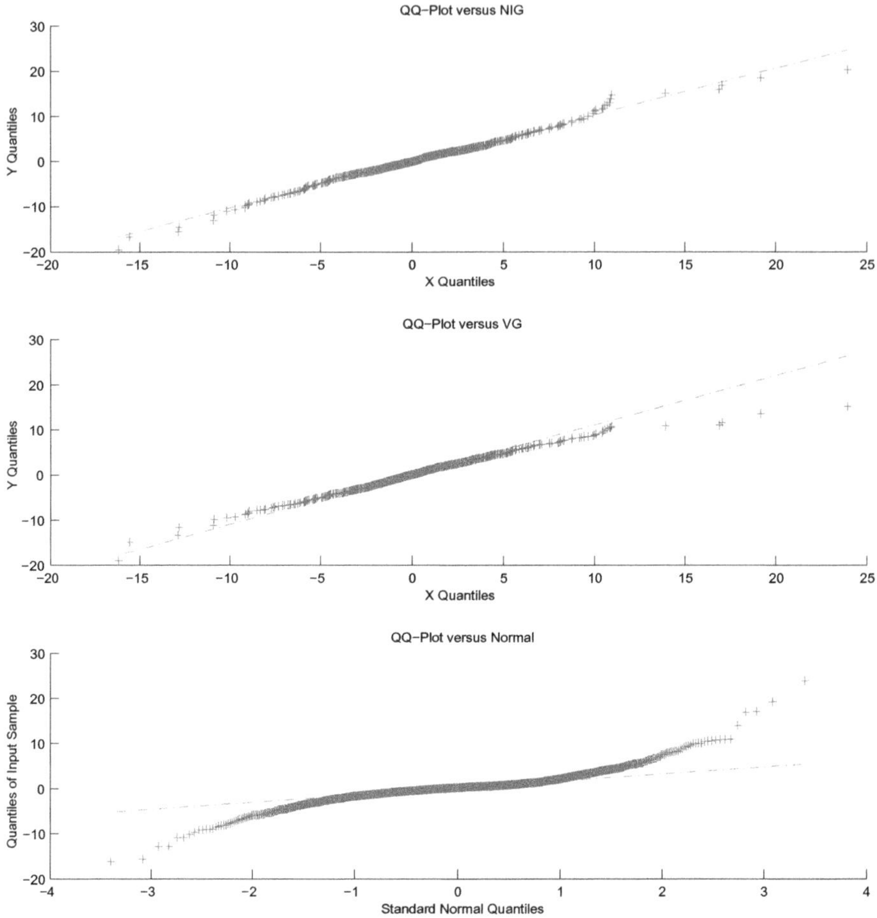

Figure 5. QQ-Plots for the Residuals of Our Lévy Model against the NIG (top), VG (middle) and the Normal (bottom) Distributed Quantiles.

where $1(\Delta z_i \leq \Delta z)$ is the indicator function equals to 1 if $\Delta z_i \leq \Delta z$, and 0 otherwise.

The Kolmogorov-Smirnov statistic for a given cumulative distribution function $F(\Delta z)$ is

$$D_n = \sup_z |F_n(\Delta z) - F(\Delta z)|,$$

where $\sup_z$ denotes the supremum with respect to the parameter z.

Table 6. Values of the Kolmogorov-Smirnov Test Statistic for Both NIG, VG and Normal Distributions.

Statistic	NIG	VG	Normal
D_n	0.0333 (0.1603)	0.0468 (0.0137)	0.1511 (0.0000)

Note: In parenthesis the corresponding p-values.

Table 7. Results of the Kolmogorov-Smirnov Tests with Respect to Different Levels α.

α	NIG	VG	Normal
0.5	1	1	1
0.3	1	1	1
0.2	1	1	1
0.1	0	1	1
0.05	0	1	1
0.025	0	1	1
0.01	0	0	1
0.005	0	0	1
0.001	0	0	1

These values can be found in Table 6. According to the Kolmogorov-Smirnov test, the null hypothesis of equality between the empirical distribution probability of our model and a reference probability is rejected at level α if

$$\sqrt{n}D_n > K_\alpha$$

where K_α is found from

$$P(K \leq K_\alpha) = 1 - \alpha.$$

K is a random variable following the Kolmogorov distribution with cumulative distribution function given by

$$P(K \leq x) = \frac{2\pi}{x}\sum_{i=1}^{\infty} \exp\left(-\frac{(2i-1)^2\pi^2}{8x^2}\right).$$

In Table 7, we reject the null hypothesis (the equality between the empirical distribution probability of our model and a reference probability, in our case the NIG distribution with the parameter from Table 2 and the corresponding Normal distribution) at the $\alpha\%$ significance level if the result is 1. Thus, our null hypothesis that the empirical data residuals follow a Normal distribution is rejected for all levels α. As for the NIG distribution, we accept the null hypothesis for all level α expected for a level of 50%, 30% and 20%. We remark again that the NIG distribution gives better results that the VG distribution, since we reject this distribution for all levels α except for the last three ones. Moreover, if we take the classical significance level α which is 5%, we reject the null hypothesis of equality between the empirical distribution probability of our model and the VG distribution probability, but we do not reject it with the NIG distribution case.

Hence, we are able to confirm that the NIG distribution provides overall a better fit to the fuel-switching price than the Gaussian distribution. Moreover, this distribution gives better results than other HG, and especially the Variance Gamma one.

5.2.3. *Cramer-von Mises goodness-of-fit hypothesis test*

Last but not least, we provide a two-sample Cramer-von Mises goodness-of-fit hypothesis test (see Anderson [1962]). This test determines if independent random samples are drawn from the same underlying distribution. The decision rules are the same as for the Kolmogorov-Smirnov test. When we applied to our model, this test yields to the results reproduced in Table 8.

We reach the same conclusions as for the Kolmogorov-Smirnov test. Indeed, we reject the both null hypotheses, with a significance

Table 8. Values of the Cramer-von Mises Test Statistic for Both NIG, VG and Normal Distributions.

Statistic	NIG	VG	Normal
Stats	0.1957 (0.7239)	0.5112 (0.9624)	6.2542 (1)
H	0	1	1

Note: In parenthesis the corresponding p-values.

level of 5%, of the equality between the empirical distribution probability of our model and the VG distribution probability. The same result applies for the Gaussian distribution. However, we do not reject the null hypothesis in the NIG distribution case. Note that similar results can obtained by resorting to Chi-squared goodness-of-fit tests.[16]

6. Conclusion

This chapter enhances our understanding of emissions abatement channelling through power systems that may be achieved in cap-and-trade schemes such as the EU ETS. Interactions between power and carbon markets occur through the fuel-switching mechanism. In this chapter, we provide empirical evidence that a Lévy-type jump model offers satisfactory results to model the fuel-switching price, taking into account carbon emissions. Besides, we demonstrate the superiority of jumps models over the standard Brownian motion assumption. In terms of the appropriate distribution required to model the fuel-switching price, the NIG or VG distributions stand out as the best candidates, outperforming by far the normality assumption. The presence of heavy tails is also clearly demonstrated throughout the empirical analysis (especially in the QQ-plots), when contrasting the probability distribution function of the fuel-switching price with the Gaussian assumption. When evaluating the relative performance of jump processes, we reach the global conclusion that the NIG distribution provides the best goodness-of-fit to the distribution of the fuel-switching price. Taken together, these key features — in sharp contrast with previous studies (e.g., Carmona *et al.* [2009, 2010]; Carmona and Hinz [2011]) — constitute innovative findings for risk managers in the power sector.

Areas for future research include, but are not limited to: (i) the valuation of derivatives products written on the fuel switching price taking into account the possible presence of jumps; and (ii) the risk-hedging of allowances in the CO_2 markets.

---

[16]Not reproduced to conserve space, but accessible upon request to the authors.

Acknowledgments: For useful comments and suggestions on previous drafts, we wish to thank Erik Delarue (KU Leuven), Duc Khong Nguyen, as well as participants at the 2014 International Symposium on Energy and Finance Issues (ISEFI) from the IPAG Business School, Paris.

References

Alberola, E. and Chevallier, J. (2009). European Carbon Prices and Banking Restrictions: Evidence from Phase I (2005–2007). *The Energy Journal* 30(3), 51–80.

Alberola, E., J. Chevallier and B. Cheze. 2008. Price Drivers and Structural Breaks in European Carbon Prices 2005–2007. *Energy Policy* 36(2), 787–797.

Alberola, E., J. Chevallier and B. Cheze. 2009a. The EU Emissions Trading Scheme: The Effects of Industrial Production and CO_2 Emissions on European Carbon Prices. *International Economics* 116, 93–126.

Alberola, E., J. Chevallier and B. Cheze. 2009b. Emissions Compliances and Carbon Prices under the EU ETS: A Country Specific Analysis of Industrial Sectors. *Journal of Policy Modeling* 31(3), 446–462.

Anderson, T. W. 1962. On the Distribution of the Two Sample Cramer-von Mises Criterion. *The Annals of Mathematical Statistics* 33(3), 1148–1159.

Arouri, M. E. H., F. Jawadi and D. K. Nguyen. 2012. Nonlinearities in Carbon Spot-Futures Price Relationships during Phase II of the EU ETS. *Economic Modelling* 29(3), 884–892.

Ballotta, L. 2005. A Lévy Process-Based Framework for the Fair Valuation of Participating Life Insurance Contracts. *Insurance: Mathematics and Economics* 37, 173–196.

Barndorff-Nielsen, O. E. and C. Halgreen. 1977. Infinite Divisibility of the Hyperbolisch and Generalized Inverse Gaussian Distributions. *Zeitschrift fur Wahrscheinlichkeitstheorie und verwandte Gebiete* 38, 309–312.

Borovkov, K., G. Decrouez and J. Hinz. 2011. Jump-Diffusion Modeling in Emission Markets. *Stochastic Models* 27, 50–76.

Bunn, D. W. and C. Fezzi. 2009. *A Vector Error Correction Model of the Interactions Among Gas, Electricity and Carbon Prices: An Application to the Cases of Germany and the United Kingdom.* In Markets for Carbon and Power Pricing in Europe: Theoretical Issues and Empirical Analyses, F. Gulli (ed.), Edward Elgar Publishing, pp. 145–159.

Carmona, R., M. Fehr and J. Hinz. 2009. Optimal Stochastic Control and Carbon Price Formation. *SIAM Journal on Control and Optimization* 48, 2168–2190.

Carmona, R., M. Fehr, J. Hinz and A. Porchet. 2010. Market Designs for Emissions Trading Schemes. *SIAM Review* 52, 403–452.

Carmona, R. and J. Hinz. 2011. Risk-Neutral Models for Emission Allowance Prices and Option Valuation. *Management Science* 57, 1453–1468.

Çetin, U. and M. Verschuere. 2009. Pricing and Hedging in Carbon Emissions Markets. *International Journal of Theoretical and Applied Finance* 12, 949–967.

Chen, Y., J. Sijm, B. F. Hobbs and W. Lise. 2008. Implications of CO_2 Emissions Trading for Short-Run Electricity Market Outcomes in Northwest Europe. *Journal of Regulatory Economics* 34(3), 251–281.

Chesney, M. and L. Taschini. 2012. The Endogenous Price Dynamics of Emission Allowance and an Application to CO2 Option Pricing. *Applied Mathematical Finance*, forthcoming.

Chevallier, J. 2009. Carbon Futures and Macroeconomic Risk Factors: A View from the EU ETS. *Energy Economics* 31(4), 614–625.

Chevallier, J. and S. Goutte. 2015. Estimation of Lévy-Driven Ornstein-Uhlenbeck processes: Application to Modeling of CO_2 and Fuel-Switching. *Annals of Operations Research*, forthcoming.

Chevallier, J., F. Ielpo and L. Mercier. 2009. Risk Aversion and Institutional Information Disclosure on the European Carbon Market: A Case-Study of the 2006 Compliance Event. *Energy Policy* 37(1), 15–28.

Chevallier, J. and B. Sévi. 2013. On the Stochastic Properties of Carbon Futures Prices, *Environmental and Resource Economics*, forthcoming.

Cont, R. and P. Tankov. 2004. *Financial Modelling with Jump Processes.* Boca Raton, FL: Chapman & Hall/CRC.

Convery, F. J. and L. Redmond. 2007. Market and Price Developments in the European Union Emissions Trading Scheme. *Review of Environmental Economics and Policy* 1(1), 88–111.

Creti, A., P. A. Jouvet and V. Mignon. 2012. Carbon Price Drivers: Phase I versus Phase II Equilibrium? *Energy Economics* 34, 327–334.

Cvitanić, J., V. Polimenis and F. Zapatero. 2008. Optimal Portfolio Allocation with Higher Moments. *Annals of Finance* 4, 1–28.

Delarue, E. and W. D'haeseleer. 2007. Price Determination of ETS Allowances through the Switching Level of Coal and Gas in the Power Sector. *International Journal of Energy Research* 31, 1001–1015.

Delarue, E. and W. D'haeseleer. 2008. Greenhouse Gas Emission Reduction-Potential by Means of Fuel-Switching in Electricity Generation: Addressing the Potentials. *Energy Conversion and Management* 49, 843–853.

Delarue, E., A. D. Ellerman and W. D. D'haeseleer. 2009. Short-Term CO_2 Abatement in the European Power Sector. *Climate Change Economics* 1(2), 1–21.

Delarue, E., K. Voorspools and W. D'haeseleer. 2008. Fuel Switching in the Electricity Sector under the EU ETS: Review and Prospective. *Journal of Energy Engineering ASCE* 134, 4–46.

Delarue, E. and W. D'haeseleer. 2007. Price Determination of ETS Allowances through the Switching Level of Coal and Gas in the Power Sector. *International Journal of Energy Research* 31, 1001–1015.

Ellerman, A. D. and B. K. Buchner. 2007. The European Union Emissions Trading Scheme: Origins, Allocation, and Early Results, *Review of Environmental Economics and Policy* 1, 66–87.

Ellerman, A.D. and B. K. Buchner. 2008. Over-Allocation or Abatement? A Preliminary Analysis of the EU ETS Based on the 2005-06 Emissions Data. *Environmental and Resource Economics* 41(2), 267–287.

Ellerman, A. D. and S. Feilhauer. 2008. *A Top-Down and Bottom-Up Look at Emissions Abatement in Germany in Response to the EU ETS*. Working Paper #2008-17. MIT CEEPR.

Falbo, P., D. Felletti and S. Stefani. 2013. Free EUAs and Fuel Switching. *Energy Economics* 35, 14–21.

Hinz, J. and A. Novikov. 2010. On Fair Pricing of Emission-Related Derivatives. *Bernoulli* 16 (4), 1240–1261.

Jing, B.-Y., X.-B. Kong and Z. Liu. 2012. Modeling High-Frequency Financial Data by Pure Jump Processes. *Annals of Statistics* 40, 759–784.

Kassberg, S., R. Kiesel and T. Liebmann. 2008. Fair Valuation of Insurance Contracts under Lévy Process Specifications. *Insurance: Mathematics and Economics* 42, 419–433.

Liu, J., F. Longstaff and J. Pan. 2003. Dynamic Asset Allocation with Event Risk. *Journal of Finance* 58, 231–259.

Lujan, E., E. Delarue, J. Chevallier, W. D'haeseleer. 2011. CO_2 Abatement Opportunity in the UK through Fuel-Switching under the EU ETS. (2005–2008): Evidence from the E-Simulate Model. *International Journal of Global Energy Issues* 35 (2/3/4), 178–214.

Madan, D. B., P. Carr and P. Chang. 1998. The Variance Gamma Process and Option Pricing. *European Finance Review* 2, 79–105.

Martzoukos, S. H. and L. Trigeorgis. 2002. Real (Investment) Options with Multiple Sources of Rare Events. *European Journal of Operational Research* 136, 696–706.

McGuinness, M. and A. D. Ellerman. 2008. CO_2 *Abatement in the UK Power Sector: Evidence from the EU ETS Trial Period*. Working Paper #2008-10. MIT CEEPR.

Pettersson, F., P. Soderholm and R. Lundmark. 2012. Fuel Switching and Climate Energy Policies in the European Power Generation Sector: A Generalized Leontief Model. *Energy Economics* 34, 1064–1073.

Schwartz, E. S. 1997. The Stochastic Behavior of Commodity Prices: Implications for Valuation and Hedging. *Journal of Finance* 52, 923–973.

Schwartz, E. S. and J. E. Smith. 2000. Short-Term Variations and Long-Term Dynamics in Commodity Prices. *Management Science* 46, 893–911.

Seifert, J., M. Uhrig-Homburg and M. Wagner. 2008. Dynamic Behavior of CO_2 Prices. *Journal of Environmental Economics and Management* 56, 180–194.

Sijm, J., S. Bakker, Y. Chen, H. W. Harmsen and W. Lise. 2005. *CO_2 Price Dynamics: The Implications of EU Emissions Trading for the Price of Electricity*. ECN Working Paper #05-081. The Netherlands.

Sijm, J., S. J. Hers, W. Lise and B. Wetzelaer. 2008. *The Impact of the EU ETS on Electricity Prices: Final Report to DG Environment of the European Commission*. ECN Working Paper #08-007.

World Bank. 2012. State and Trends of the Carbon Market 2012. Report. Carbon Finance Unit. Washington DC.

Zachmann, G. 2013. A Stochastic Fuel Switching Model for Electricity Prices. *Energy Economics* 35, 5–13.

Zhu, Z., P. Graham, L. Reedman and T. Lo. 2009. A Scenario-Based Integrated Approach for Modeling Carbon Price Risk. *Decisions in Economics and Finance* 32, 35–48.

Chapter 14

The Use of Real Option Approach for the Evaluation of Climate Investments

Luis Mari Abadie

Basque Centre for Climate Change (BC3),
Edificio Sede N°, Planta 1ª
Parque Científico de UPV-EHU. Leioa (Bizkaia)
lm.abadie@bc3research.org

Ibon Galarraga

Basque Centre for Climate Change (BC3),
Edificio Sede N°, Planta 1ª
Parque Científico de UPV-EHU. Leioa (Bizkaia)
ibon.galarraga@bc3research.org

Abstract

Climate finance refers to flows of resources or funds for activities related to combating climate change, i.e., reducing greenhouse gases (GHGs) emissions or adapting to climate change. This definition can cover many types of investment, and indeed their number is increasing in the fields of both mitigation and adaptation. Such investments entail numerous uncertainties and therefore risks, which can sometimes be reduced. In many cases, uncertainties persist to a large extent as regards to future cash flows. These investments can be assessed using financial methods that take their specific aspects into account, i.e., considering the stochastic behavior of significant variables such as the price of emission allowances, electricity prices in the case of renewables, fossil fuel prices if the project involves replacing or reducing the output of such fuels, sea level rise (SLR) in some adaptation cases, temperature increases, the risk of default on projects, exchange rate risks, uncertainty in the cost of investment and other case-specific factors.

Firms that invest in projects of this type clearly expect returns in proportion to the risks that they run.

This chapter uses a number of examples to demonstrate the application of techniques developed in the field of finance to valuing investments classified as climate finance. The chapter specifically deals with real option (RO) approach. This approach is tailored specifically to valuing irreversible investments under conditions of uncertainty and can include the case of projects with risk of default.

The examples here use market quotes for the variables included in the analysis whenever such information is available.

The results show that it is important to factor uncertainty into the valuation of almost all types of investment, but especially so in the case of climate finance investment.

1. Introduction

Climate finance refers to flows of resources or funds for activities related to combating climate change, i.e., reducing greenhouse gases (GHGs) emissions or adapting to climate change. This topic has attracted a lot of attention in climate negotiations, especially since the Cancun Accords were adopted in 2010. The Accords included a commitment to raise USD 100 billion per year to finance the mitigation and adaptation needs of developing countries. Buchner *et al.* [2014] offer a broad, detailed review of the status of climate finance flows, showing that although the global climate finance flows reached USD 331 billion by 2013, the gap between those flows and the needs of developing countries continues to increase. In addition, climate finance flows have decreased for two consecutive years. 40% of that figure of USD 331 billion came from public sources, which means that the private sector was responsible for almost 60%. The same study states that "almost three-quarters of climate finance flows were invested with the expectation of earning commercial returns". And more than 90% of investment was devoted to mitigation projects (mainly renewable energy (RE) (78%) followed by energy efficiency (10%) and sustainable transport (6%)). Out of the flow for financing adaptation (7% of the total), water management accounted for the biggest share (58%) followed by infrastructure and coastal protection (14%), disaster risk management (9%) and agriculture and forestry (8%).

Uncertainty is one of the main barriers to the effective materialization of some of these flows, and in the context of climate change there are a number of sources of uncertainty. Markandya [2014] offers a good overview of how uncertainty affects investment decisions in the field of climate change and shows some examples of methodologies to address it in project appraisal.

The Real Option (RO) approach is one of the soundest methods for dealing with investment valuations when irreversibility and uncertainty are present [Dixit and Pindyck, 1994]. One of its distinguishing features is that it can factor in the flexibility available to managers, e.g., the power to choose the best time to invest, to decide whether to withdraw from a project temporarily or permanently, to change the scale of a facility and to change inputs and outputs. This attribute becomes especially relevant in the field of adaptation and mitigation investments due to the number of uncertainties that could affect the decision (such as those deriving from climate modelling, socio-economic scenarios, prices of emission allowances and many others related to investments). Trigeorgis [1996] classifies real options according to various types: option to defer, time-to-build option, option to alter operation scale, option to abandon, option to switch, growth options and multiple interacting options. Each of these is appropriate for different types of project.

Many climate finance investment decisions in both mitigation and adaptation can also be classified as irreversible. The decisions to build a wind farm or a hydroelectric power plant are examples of almost irreversible investments in mitigation; the construction of a dam to prevent damage from sea level rise (SLR) is an example in regard to adaptation, although some flexibility can be factored in if size can be adapted in the light of updated climate information becoming available.

Moreover, uncertainty is a significant element in this type of investment, and one that appears in many aspects [Abadie and

Chamorro, 2013]:

(a) in the prices of the major variables, e.g., CO_2 emission allowances and electricity generated using renewables;
(b) in quantities, e.g., the amount of wind at a wind farm;
(c) in investment costs, especially when planning for a more or less long-term future;
(d) in climate variables such as temperature and SLR in the future;
(e) in uncertainty on regulatory issues such as changes in certain subsidies and/or rates that may affect the outcome of investments in renewable energy (RE);
(f) in financial variables such as exchange risk; and
(g) in the default risk case.

Given that investments of this type are irreversible (even though they may allow for some flexibility) and so is climate change, time and the discount factor are highly important in valuing them.

As stated earlier, renewable energy (RE) accounts for much of the investment made in mitigation. Menegaki [2008] reviews the literature on the valuation/evaluation of renewable energy resources and summarizes the methods used. The author discerns four major categories: economic welfare-based methods, financial economics-based methods, emery analysis and economic but not welfare-based methods. According to IRENA [2012], RE investment has generally followed broader trends in foreign direct investment (FDI). Investments in RE in developing countries are one of the chief mitigation measures taken, but they come up against added difficulties such as exchange rate risks (which may be difficult to cover in full), country risk and the lack of or weakness of capital markets in such countries. This heightened uncertainty can lead investors to demand higher expected returns, lower loan-to-value ratios or shorter maturity periods before they are prepared to invest in RE there. The same goes for other investments that fall under the heading of climate finance.

The expansion of RE in recent years has been largely based on support policies including initial, one-off transitory subsidies, constant feed-in tariffs, market prices plus a fixed premium, market prices plus a Renewables Obligation Certificate (ROC) price and other incentive schemes. These support policies seek to address market failures. Investments of other types may of course require different support instruments such as special loans, subsidies, etc.

Despite some difficulties in implementation in the past, in recent years renewable energy auctions have become a popular policy tool for selecting investors. The number of such auctions increased from nine in 2009 to at least 44 by early 2013, 30 of them in developing countries [IRENA, 2012].

All the major characteristics of each investment project must be taken into account in evaluating it, and those characteristics may differ significantly from one country to another. Funding for such projects may comprise different combinations of loans and equity. They may also make use of mezzanine finance.[1] Multilateral development banks (MDBs) tend to play a major role in the funding of investment projects in developing countries, as indeed do the Green Climate Fund and other investment support schemes that have been developed in the field of climate change.

In financial literature, there are many publications that deal with valuation under uncertainty and that may be useful in regard to climate finance investment both for mitigation and adaptation. Useful financial books include Hull [2011], Wilmott [2006] and Luenberger [1998]. Books dealing specifically with the RO approach include Dixit and Pindyck [1994] and Trigeorgis [1996]. The application of RO to investments in energy under uncertainty is tackled in Abadie and Chamorro [2013], Abadie and Chamorro [2008a] and Ronn [2002]. Recent examples can also be found that deal with projects of other types concerned with adaptation [Hertzler, 2007], with natural resources [Mezey and Conrad, 2010], with flood risk management [Woodward *et al.*, 2011] and with SLR [Kontogianni *et al.*, 2014]. Finally, a good review of uncertainty in the field of climate change

[1]Mezzanine financing is a hybrid between debt and equity. The lender typically has a legal right enabling him to convert the security into equity at a predetermined price per share if the loan is not paid on time or in full. Being unsecured and highly subordinated, mezzanine financing has high interest rates.

that also covers RO can be found in Hallegatte *et al.* [2012] and Watkiss *et al.* [2014].

When investments in mitigation are made in developed countries the risks are usually lower, because there are consolidated capital markets and lower country risk levels. In such countries, there are also organized markets and over-the-counter (OTC) investment schemes that provide a great deal of information and enable valuations to be drawn up using the risk-neutral method assuming the existence of complete markets,[2] as is usually done in valuing financial derivatives such as options and futures.

In those countries, there are usually markets for commodities related to climate finance, such as the EU emissions trading system (EU ETS). The information provided by such markets facilitates valuations such as that of Abadie and Chamorro [2008b], who examine an option for investing in carbon capture and storage (CCS) for a coal-fired power plant and derive the optimal investment rule. In their paper, CO_2 prices are modelled as geometric brownian motion (GBM) processes while the electricity price is modelled as a mean-reverting process. They show that uncertainty has a significant effect on investment decisions, and that a high CO_2 price is needed for investment in CCS to be worthwhile, so the best option is to wait and see whether prices change favorably. Similar work is undertaken in Abadie *et al.* [2014], in which the authors evaluate two alternative CCS technologies at a coal-fired power plant from an investor's point of view. The first technology uses CO_2 for enhanced oil recovery (EOR) paired with storage in deep saline formations (DSF) and the second merely stores CO_2 in DSF. They evaluate investment with three risk sources: the price of electricity, the price of oil and the price of carbon allowances. Another application to energy assets using the RO approach is examined in Abadie and Chamorro [2008b].

In the case of adaptation investments, RO is also a very useful technique as it enables users to analyze flexibility, learning and

[2]For complete markets, the Arbitrage Theorem states that there is no arbitrage if and only if there is a risk-neutral probability measure. This does not mean that investors are necessarily risk-neutral.

future information while offering a sound method for evaluating the appropriate timing of the investment or the adjustment in size of certain projects [Watkiss *et al.*, 2014]. A simple example would be the possibility of building a wall with deeper foundations than originally planned so it can be adapted (raised) in the future should SLR be greater than expected. In such a case the idea of "buying" robustness should be evaluated with an RO approach to adequately determine whether it is good value for money [Markandya, 2014].

This chapter is devoted to illustrating the use of RO techniques in the climate finance investments context. The rest of the chapter is structured as follows: Section 2 presents stochastic models, Section 3 discusses valuation under uncertainty and conclusions are drawn in Section 4.

2. Stochastic Models

One of the particularities of the RO approach is that it enables a shift to take place from a more traditional deterministic approach to an analysis that implicitly incorporates uncertainty. This is done with the use of stochastic models.

Stochastic models are useful for showing how any of the uncertain variables that need to be accounted for in an investment decision behave. These variables may include for instance the price of a certain input, output, emission allowance or the losses generated as a consequence of the rate of SLR.

It can be assumed that a variable S which is significant for climate finance can evolve stochastically in accordance with a number of models.

We consider the three possible cases:

The first is a geometric Brownian motion process (GBM) as per equation (1):

$$dS = \mu S dt + \sigma S dW. \tag{1}$$

The second is a mean-reverting process (MR) as per equation (2):

$$dS = k(S_m - S)dt + \sigma S dW. \tag{2}$$

The third is a mean-reverting process with jumps (MRJ) as per equation (3)[3]:

$$dS = k(S_m - S)dt + \sigma S dW + \phi S ds. \quad (3)$$

In these equations, μ is the GBM drift rate, σ is the volatility, $dW = \varepsilon(t)\sqrt{dt}$ is the increment of a Wiener process with $\varepsilon(t) : N(0, 1)$, S_m is the long-term equilibrium value towards which S tends to revert in a mean-reverting process and k is the speed of reversion that determines how the expected value of S approaches the long-term equilibrium value S_m in time. In the MRJ process, $ds = 1$ with probability λdt and $ds = 0$ with probability $1 - \lambda dt$, $\lambda > 0$ is the mean arrival rate, ϕS is the value of the jump, if any. The value of ϕ can be stochastic.

In the GBM process, the drift rate μ is the tendency of the deterministic part of equation (1), and allows calculating the expected value for a certain moment in the future. Volatility σ is the parameter that determines the risk.

In the MRJ process, parameter λ determines the expected frequency for jumps $1/\lambda$ (flood events every 10 or 50 years for instance); it is an expected value of an event that may or may not occur.

3. Valuation under Uncertainty

In an attempt to illustrate the use of RO, we present two examples here: the first is investment in a mitigation project that enables savings to be made in CO_2 emission allowances. For instance, this could be any carbon abatement project. The second looks at the risk of default from the viewpoint of the lender, and can be applied to both mitigation and adaptation contexts.

3.1. *Climate finance and the GBM model: An application to a carbon abatement project*

Once it has been decided which model is to be used to show the behavior of the variable S, in this case GBM, the following expected

[3]Note that although this is not used in the examples later in this chapter, it has been included for illustration purposes.

value can be calculated:

$$E(S_T) = S_0 e^{\mu T} \tag{4}$$

depending on whether μ is the drift in the real world or in the risk-neutral world. Assume in this case that we want to evaluate a carbon abatement project and therefore we are interested in understanding the behavior of the EU ETS emission allowance.

From equation (4), it can be derived that the value of income received between τ_1 and τ_2 when the interest rate is r is given by:

$$V(S_0) = S_0 \frac{e^{(\mu - r)\tau_2} - e^{(\mu - r)\tau_1}}{\mu - r}. \tag{5}$$

The trading prices from the futures market (from 29 December 2014 for this example) can be used to estimate the drift in the risk-neutral world (See Figure 1). The figure obtained is $\mu = 0.0216$.

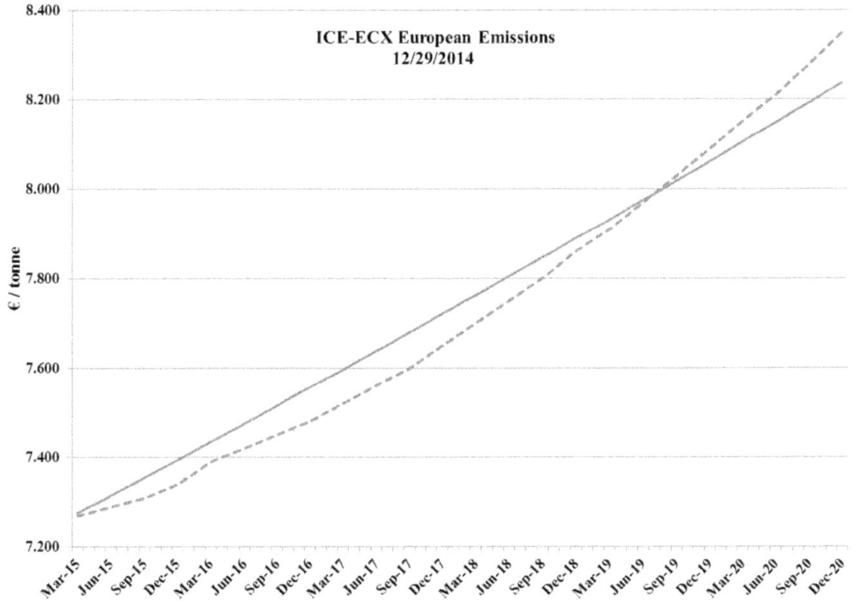

Figure 1. ICE-ECX European Emission Futures Prices (29 December 2014). In addition, the spot prices for 2014 shown in Figure 2 are used to calculate the volatility.

Figure 2. European Emissions Spot Prices.
Source: SENDECO2.

In this case the figure obtained is $\sigma = 0.4665$, so the prices of the allowances are very low and fairly volatile.

Now assume a risk-free interest rate of $r = 0.025$ and consider the case with a crude time discretisation: $T = 1$, and $\Delta t = 1$; i.e., the whole time horizon (a year) is approximated by a single time step.

Although it is possible to invest in a carbon-abating project either now or in a year's time, upon investment the savings (or sale revenues from spare allowances) will accrue over 20 years as from one year after the decision was made (it is assumed that the facility will take one year to build; $\tau_1 = 1$, $\tau_1 = 21$).

The remaining parameters are as follows: the investment cost is $I =$ €110 and the initial spot price is $S_0 =$ €7.25/tonne.

We construct a binomial tree (Figure 3) to carry out the valuation.

The lattice of the asset price aims to match the probability distribution in continuous time as closely as possible. The values of u, d, and p are determined in such a way that this is accomplished. In

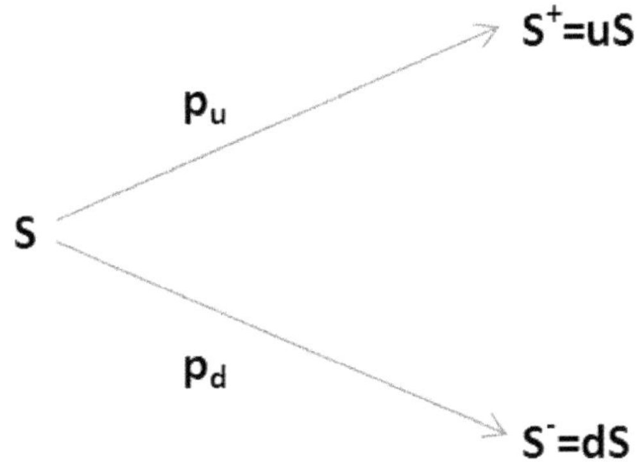

Figure 3. One Step Binomial Tree.

particular, there must be a strict correspondence regarding the mean (or mathematical expectation) and also the variance over any short time interval Δt.

Concerning the former the following must hold:

$$Se^{\mu\Delta t} - uSp_u + (1 - p_u)dS \leftrightarrow e^{\mu\Delta t} - p_u u + (1 - p_u)d,$$

with $p_u = \frac{e^{\mu\Delta t} - d}{u - d} = 0.4080$.

As for the latter using $d = 1/u$ gives:

$$u = e^{\sigma\sqrt{\Delta t}}, \quad d = e^{-\sigma\sqrt{\Delta t}}.$$

The decision to invest at the end is assessed first, then that of investing at the beginning.

After one period, the emission allowance price can go either up or down:

$$S^+ \equiv uS = S_0 e^{\sigma\sqrt{\Delta t}} = 11.5594, \quad S^- \equiv dS = S_0 e^{-\sigma\sqrt{\Delta t}} = 4.5472.$$

If the investment is made after one year, the present value of a 20-year income is:

$$V(S_0) = 19.2695 S_0.$$

Thus, if investment takes place at the expiry of the option one of these two payoffs is received:

$$V(11.5594) - 110 = 112.74, \quad V(4.5472) - 110 = -22.38.$$

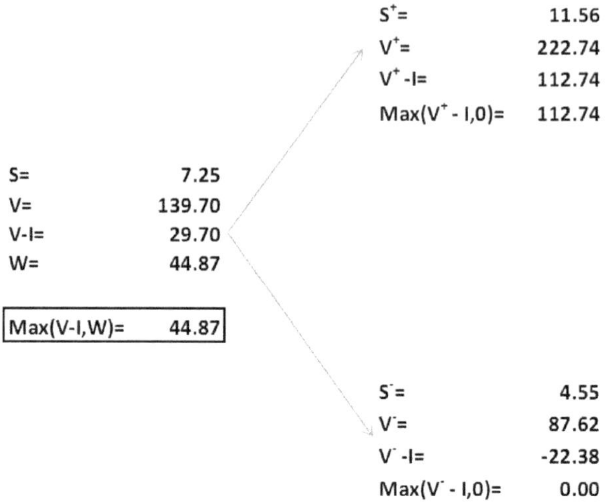

Figure 4. One-Step Binomial Tree Example (Baseline Case).

And the investment decisions will be as follows:

$$Max(V(11.5594) - 110.0) = 112.74, \quad Max(V(4.5472) - 110.0) = 0.$$

At the initial time:

Investment Value $= V(7.25) - 110 =$ €29.70

Continuation Value $= e^{-r\Delta t}(p_u W^+ + p_d W^-) =$ €48.87

W $=$ max (Invest, Continue) $=$ €48.87.

Thus, in this case it is preferable to wait at the outset because the continuation value is greater than the investment value. These results are shown in Figure 4 and confirm that even if the Investment Value (or Net Present Value, NPV) at the initial time is positive, it is not optimal to invest but rather to wait and see due to the uncertainty expressed by σ.

When more steps are used, the resulting model is similar but the results are more accurate: on the one hand the stochastic process is represented better, and on the other hand more decision-making times are admitted in the cases of American (i.e., the option can be exercised any time before the end of the period) or Bermudan

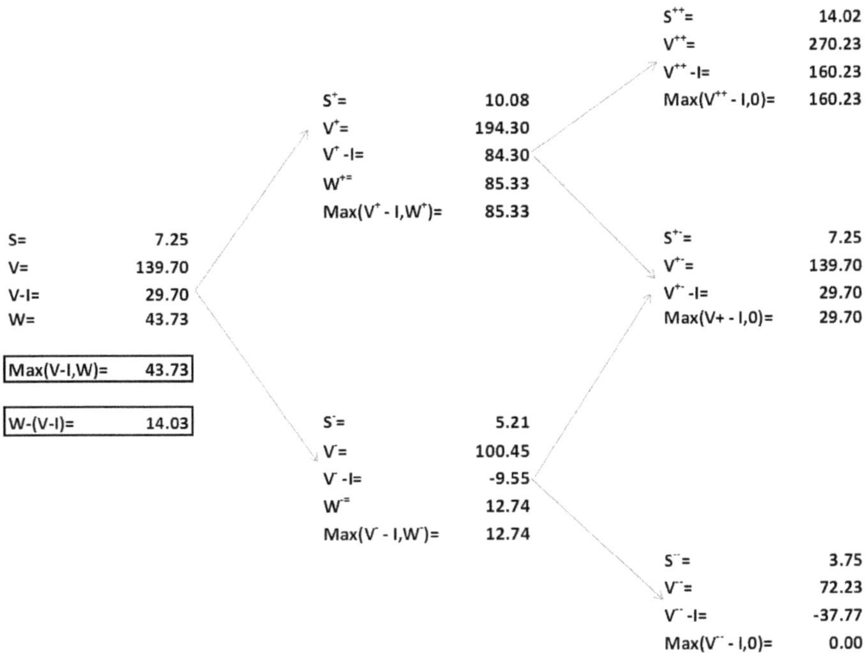

Figure 5. Two-Step Binomial Tree Example (Baseline Case).

options (i.e., the option can only be exercised in a given number of times). For the sake of illustration, this impact is shown in Figure 5 with two steps per year. As can be seen, this gives a continuation value of 43.73, somewhat lower than the 44.87 obtained with one step per year.

Given $S_0 = 7.25$, it is now possible to look for the threshold investment cost I^* below which it is optimal to invest immediately.

This is indeed the best course of action when the NPV exceeds the continuation value. Note, though, that both parts of the comparison depend on I. Figure 6 shows that the two values are the same for $I^* = 19.2055$.[4] Below I^* investment is made immediately and above it the best course is to wait and take the best decision in a year's time.

[4]Note that this value has been obtained by changing I. So V-I = W.

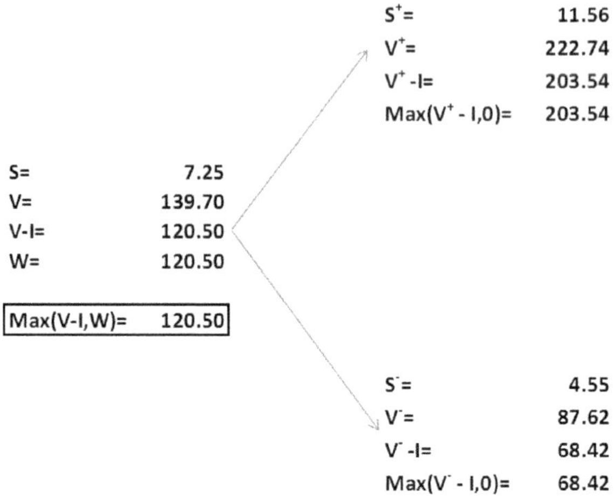

Figure 6. Threshold Investment Cost (Case II).

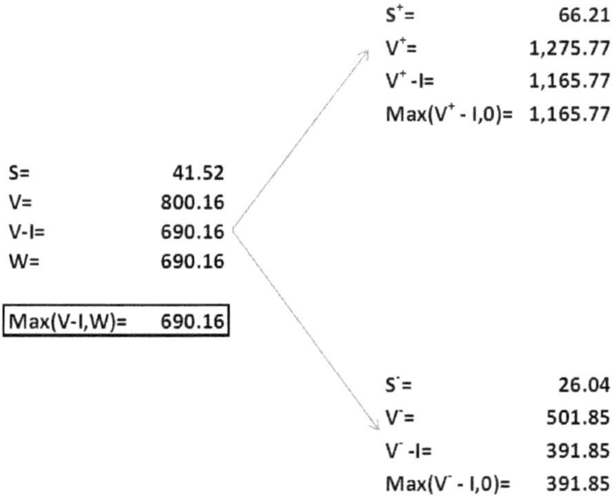

Figure 7. Threshold Allowance Price (Case III).

Therefore, in a real application of this model it becomes very important to choose a suitable number of steps to accurately represent the stochastic behavior.

This raises the following question: given $I = 110$, what is the trigger carbon price S^* above which it is optimal to invest immediately? As before, the two values to be compared depend on S. The critical price turns out to be $S^* = 41.52$. This result is shown in Figure 7.

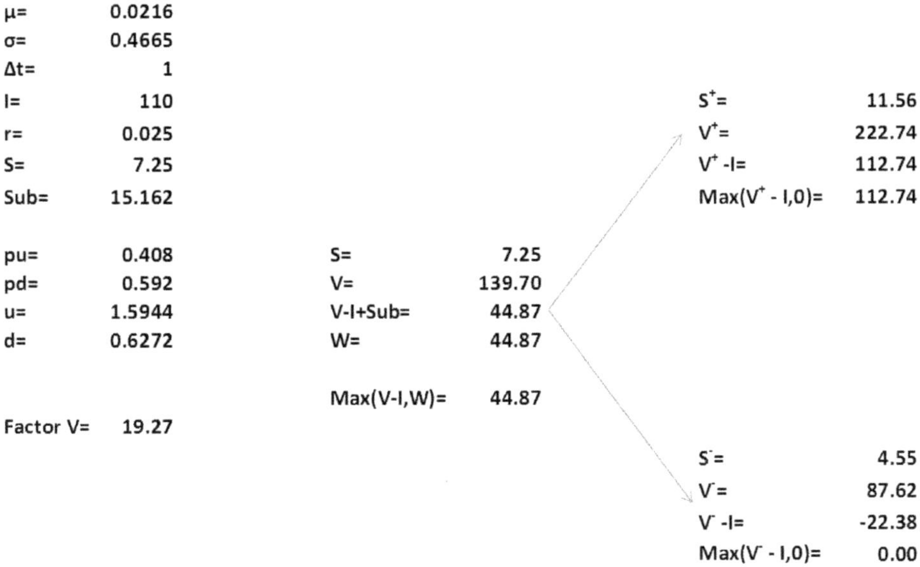

Figure 8. Binomial Tree with Subsidy (Case IV).

Results show that as a consequence of uncertainty a CO_2 price of €41.52 per tonne is required for the decision to be to invest now.

Now assume that the intention is to support this type of investment with a subsidy at the outset that will not be available in a year's time. Figure 8 shows that investment is immediate with subsidies reaching €15.162 per tonne and more.

As can be seen, it is now preferable to invest at the outset in spite of the uncertainty.

All these cases are illustrative in that they use $\Delta t = 1$, which gives rise to some degree of inaccuracy because it does not properly represent the stochastic GBM process. It would be better to use more steps: e.g., $\Delta t = 0.1$ and $\Delta t = 0.01$ give the results shown in Table 1.

Although the NPV and the continuation value are very similar in both cases, the boundary values are not. It can be shown that a subsidy at the outset only is a highly effective way of significantly promoting these investments.

Increasing the number of steps improves the fit of the binomial tree to the stochastic process in continuous time and also enables the

Table 1. Comparison of Results Depending on n° of Steps.

	Δt = 1	Δt = 0.1	Δt = 0.01
Baseline Case			
. NPV	29.70	29.70	29.70
. Continuation Value	44.87	42.19	42.12
Case II			
. I*	19.21	15.30	14.56
Case III			
. S*	41.52	52.13	54.90
Case IV			
. Subsidy	15.16	12.48	12.42

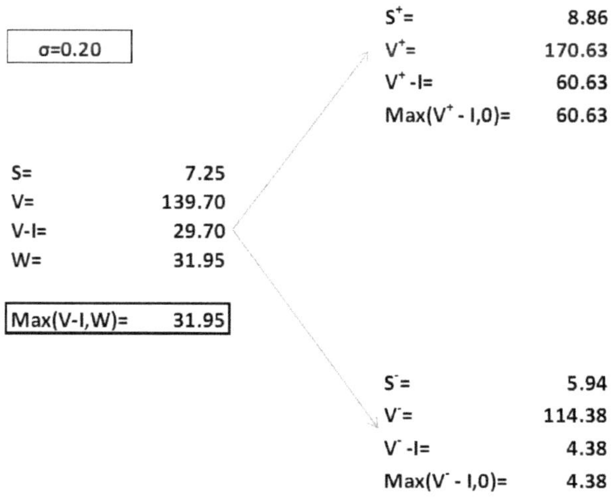

Figure 9. Binomial Tree with Lower Volatility (Case V).

option to be exercised at interim times. With more steps, what we are really doing is valuing an American option better.

Volatility is a significant element in valuation under uncertainty. Figure 9 shows the impact of a reduction in volatility: the continuation value is markedly lower.

3.2. *Project finance and climate finance: Decisions under the risk of default*

Some climate finance projects — be they in the field of mitigation or adaptation — may involve major uncertainties and there may be a risk of default for the lender. Winsen [2010] presents a model based on GBM-type stochastic behavior in the project value to analyze the risk of default.

The approach used below is similar but more highly developed. We assume that the underlying variable S follows a mean-reverting stochastic process (described in Section 2) and that the value of the project is a function of that variable, i.e., $F(S)$[5] can be financed with non-recourse loans with collateral for the project value $F(S)$, i.e., a function of a particular price such as, for instance, the price of electricity. In this example, it is assumed that S follows a mean-reverting process as per equation (2).

At the final time T, it is possible that the value of the Project $F(S_T)$ may be lower than L_T, the amount of the loan outstanding at T. If so there is a default, and if not then the loan is paid off, so the final value at time T is:

$$V_{Tv} = Min(F(S_T), L_T). \tag{6}$$

There is the initial value of the Project $F(S_0)$ and a loan L_0 at a constant interest rate of i. Initially the loan-to-value ratio is:

$$LTV = L_0/F(S_0). \tag{7}$$

The expected value $E(S_t)$ at a future time t is given by:

$$E(S_t) = S_m + [S_0 - S_m]e^{-kt}. \tag{8}$$

This type of mean-reverting model is analyzed in Abadie and Chamorro [2013] for the case of Energy Investments.

[5]In Winsen [2010], $F(S) = S$ with S behaving as per a GBM stochastic model.

The value of an annuity at T is therefore as follows:

$$F(S_T) = \int_{\tau_1}^{\tau_2} E(S_t)e^{-rt}dt = \frac{S_m}{r}[e^{-r\tau_1} - e^{-r\tau_2}] + \frac{S_0 - S_m}{k+r}[e^{-(k+r)\tau_1} - e^{-(k+r)\tau_2}]. \tag{9}$$

Now consider a case with these parameter values: $S_0 = 80$, $S_m = 100$, $\sigma = 0.25$, $r = 0.05$ and $k = 0.20$. For the sake of illustration, consider $\Delta t = 1$ year, though in real life it is better to divide time into smaller steps. In one period, the underlying price S may increase to:

$$uS = S^+ = S_0 e^{\sigma\sqrt{\Delta t}} = 102.72 \quad \text{or decrease to } dS = S^- = S_0 e^{-\sigma\sqrt{\Delta t}} = 62.30.$$

Denote $\Delta X = \sigma\sqrt{\Delta t}$. This gives $\eta = \frac{k(S_m - S_0)}{S_0} - \frac{\sigma^2}{2} = 0.01875$ and the probabilities are

$$p_u = \frac{1}{2} + \frac{\eta\Delta t}{2\Delta X} = 0.5375 \; p_d = 1 - p_u = 0.4625. \tag{10}$$

If the project produces one unit per annum from the outset and for 20 years the amount of immediate investment in the project value can be calculated using equation (9):

At the outset $F(S_0) = €1,184.78$ and at the end $F(S^+) = €1,275.06$ and $F(S^-) = €1,114.47$.

Assume a loan $L_0 = €1,100$ with an interest rate of $i = 0.09$ if the loan and the interest are paid off at the end of the first year, $L_0 = €1,203.59$. In these circumstances, the possible final values for the lender would be €1,203.59 with a likelihood of 53.75% and €1,114.47 with a likelihood of 46.25%, which gives an average of €1,162.37. Converted to present value with $r = 0.05$ this works out to €1,105.68. As shown in Figure 10, this gives an expected project finance value greater than the loan provided.

A drop in the interest rate i to i = 0.0807 would leave the expected value equal to the face value of the loan. This result is shown in Figure 11.

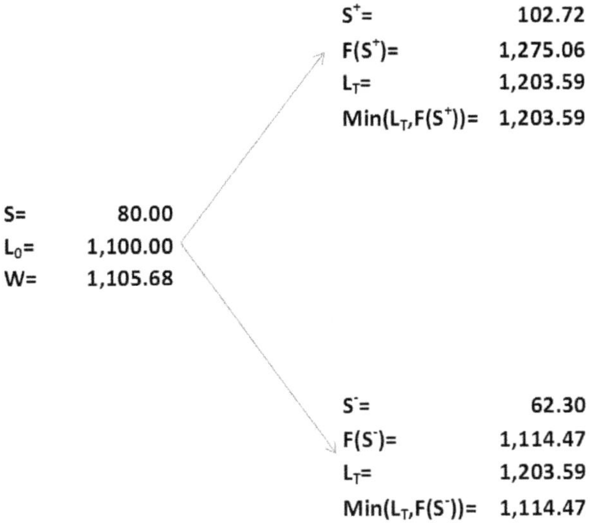

Figure 10. Baseline Case Project Finance.

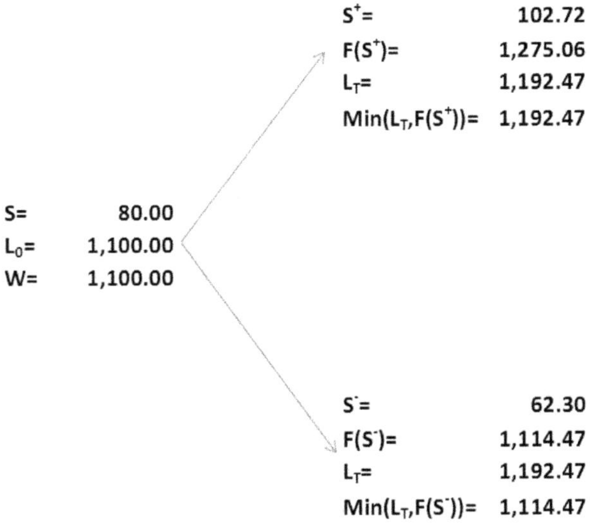

Figure 11. Case with Interest on Loan i = 8.07%.

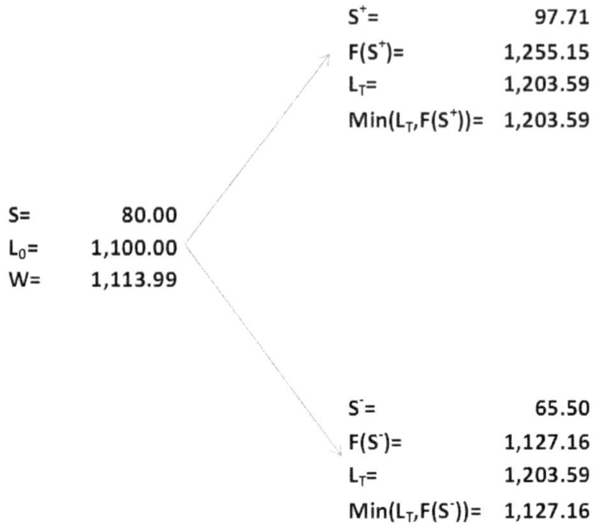

Figure 12. Case with Volatility = 20%.

The impact of a drop in volatility can also be analyzed. If $\sigma = 0.20$, the results shown in Figure 12 are obtained. As can be seen, in this case lower volatility makes investment more attractive.

Keeping the rest of the parameters constant, it can be calculated that investment is only possible if volatility is below 0.2849. These results are shown in Figure 13.

When there is a default there are usually default costs (df), so the amount recovered is lower. Taking df = 50 the results are as shown in Figure 14, where it can be seen that investment is not attractive. The higher the default costs are for a given project in a given country, the less attractive financing the investment is to potential lenders.

This example can be extended in various ways, such as more steps per year, other forms of repayment and other interest payment arrangements.

Finally, consider a representation with a better fit, with 100 steps per year but with interest being paid and the loan capital being paid off, as the case may be, only at the end of the year, so that the possibility of default exists only at the end. In this case, the current value of the project is W = 1,081.9.

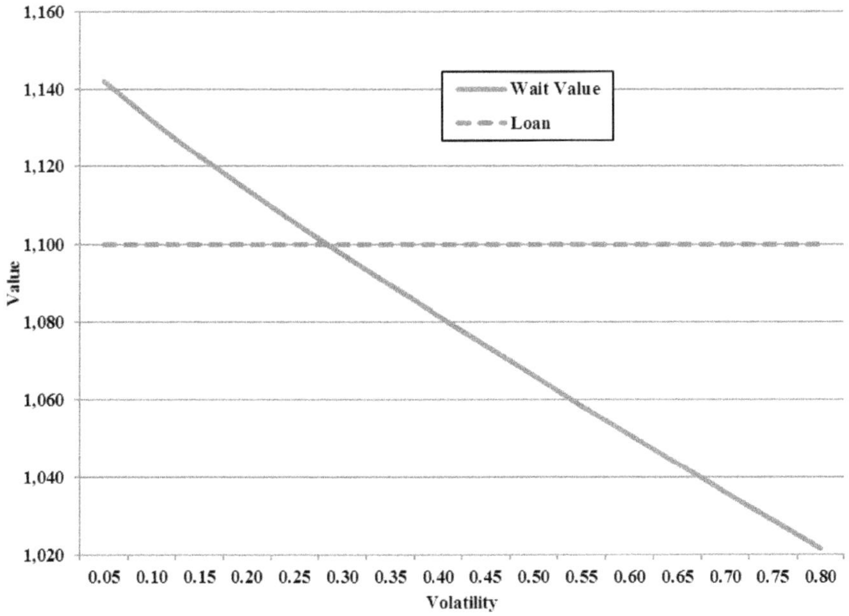

Figure 13. Project Value and Volatility.

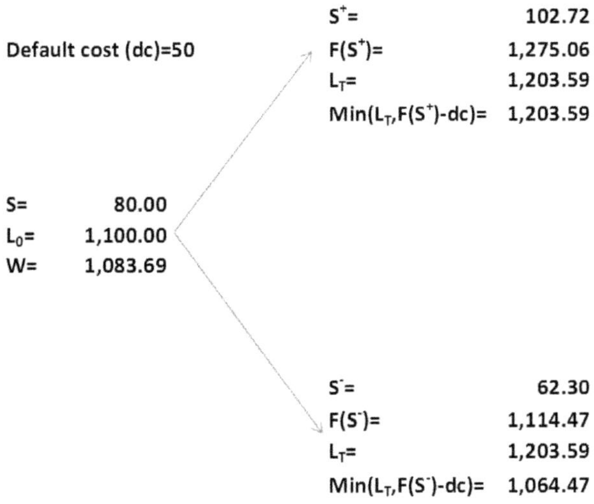

Figure 14. Case with Default Cost = 50.

4. Conclusions

This chapter shows how investments in climate finance can be valued under uncertainty using the RO approach. Uncertainty shows up as a significant element in valuations concerned with both mitigation and adaptation projects. Failure to take uncertainty into account may result in highly costly wrong decisions. Moving from a deterministic approach such the ones resulting from Cost Benefit Analysis to an RO approach may result in a completely different outcome. The latter considers the value of the "option" to do something else (abandon investment, re-adjust the size of the project, etc.) at some point in the near future. The option gives the right but not the obligation to take the alternative course.

In this context, it is important to select the main sources of risk for each investment project and to correctly model them stochastically. For the sake of illustration, most models are shown here with a single step per year, but it is pointed out that for proper adaptation to the stochastic model and for flexibility in decision-making smaller steps should be used in the binomial decision-making trees. That is, the models should account for the possibility that decisions can be made at other times. The calculation of optimal decisions on investment under uncertainty is also illustrated here. The methodologies shown here are a very interesting contribution to guiding investment decisions.

Two sets of examples are given: the first concerned with investment in mitigation, enabling savings to be made in certain CO_2 emission allowances that have financial value. This could be the representation of any mitigation project in which the only source of uncertainty is the price of the allowances under the EU ETS. In this case, the method selected is risk-neutral valuation using market data.

The second set looks at the risk of default from the viewpoint of the lender, and can be applied to both mitigation and adaptation. This could be illustrated for instance by any project in which the developer commits to repay part of a loan received from a multilateral climate fund. In this case, a real-world valuation is used, with actual probabilities and a discount rate that takes the risk into account.

The use of an RO approach for climate finance investment is tantamount to incorporating the value of waiting, buying flexibility or even abandoning investments in the light of uncertainties. And of course, if one of the goals of climate policies is to promote investment in climate finance it becomes a requisite that financial aspects that may condition such investments be taken into account.

Some of the lessons learnt in this chapter are that choosing the right time to invest is really important and that methodologies that help in this process are essential in climate finance. In fact, some of the traditional tools may be misleading under some circumstances: for instance, we show that in certain cases projects with a positive NPV might not be worth undertaking until the value of certain variables reaches certain levels, i.e., trigger prices. In this context, concepts such as volatility play a significant role, showing that uncertainty often leads to investment decisions being delayed, especially in the case of investments that are irreversible.

We argue that stochastic modelling offers much better guidance for investment decision making and choosing the right number of moments or steps becomes a crucial issue. If policy instruments are to be used to support investments in climate finance then using a subsidy at the outset may be considered. This is also illustrated in this chapter. Finally, when there is a risk of default, the loan-to-value rate, default costs and volatility measures to reduce the risk can be very effective.

Acknowledgments: Luis Mari Abadie and Ibon Galarraga gratefully acknowledge financial support received from the Basque Government via project GIC12/177-IT-399-13 and from the Research Council of Norway under the project "Strategic Challenges in International Climate and Energy Policy (CICEP)" (http://www.cicep.uio.no/english/).

References

Abadie, L. M. and J. M. Chamorro. 2008a. Valuing Flexibility: The Case of an Integrated Gasification Combined Cycle Power Plant. *Energy Economics* 30(4), 1850–1881.

Abadie, L. M. and J. M. Chamorro. 2008b. European CO_2 Prices and Carbon Capture Investments. *Energy Economics* 30, 2992–3015.

Abadie, L. M. and J. M. Chamorro. 2013. *Investment in Energy Assets under Uncertainty*. London, UK: Springer-Verlag.

Abadie, L. M., I. Galarraga and D. Rübbelke. 2014. Evaluation of Two Alternative Carbon Capture and Storage Technologies: A Stochastic Model. *Environmental Modelling & Software* 54, 182–195.

Buchner, B., M. Stadelmann, J. Wilkinson, F. Mazza, A. Rosenberg and D. Abramskiehn. 2014. *The Global Landscape of Climate Finance 2014*. Climate Policy Initiative.

Dixit, A. K. and R. S. Pindyck. 1994. *Investment under Uncertainty*. Princeton, NJ: Princeton University Press.

Hallegatte, S., A. Shah, R. Lempert, C. Brown and S. Gill. 2012. *Investment Decision Making under Deep Uncertainty Application to Climate Change*. Policy Research Working Paper 6193. The World Bank.

Hertzler, G. 2007. Adapting to Climate Change and Managing Climate Risks by Using Real Options. *Australian Journal of Agricultural Research* 58, 985–992.

Hull, J. C. 2011. *Options, Futures and Other Derivatives*. 8th edn. Upper Saddle River, NJ: Prentice Hall.

IRENA. 2012. Financial Mechanisms and Investment Frameworks for Renewables in Developing Countries. [Internet]; [updated 2015 January 9]. At: http://irena.org/Finance_RE_Developing_Countries.pdf.

Kontogianni, A., C. H. Tourkolias, D. Damigos and M. Skourtos. 2014. Assessing Sea Level Rise Costs and Adaptation Benefits under Uncertainty in Greece. *Environmental Science and Policy* 37, 61–78.

Luenberger, D. G. 1998. *Investment Science*. New York, NY: Oxford University Press.

Markandya, A. 2014. *Incorporating Climate Change into Adaptation Programmes and Project Appraisal: Strategies for Uncertainty*. In Routledge Handbook of the Economics of Climate Change Adaptation, A. Markandya, I. Galarraga and E. Sainz de Murieta (eds.), Oxon, UK and New York, NY: Routledge.

Menegaki, A. 2008. Valuation for Renewable Energy: A Comparative Review. *Renewable and Sustainable Energy Reviews* 12, 2422–2437.

Mezey, E. W. and J. M. Conrad. 2010. *Real Options in Resource Economics. Annual Review of Resource Economics* 2, 33–52.

Ronn, E., ed. 2002. *Real Options and Energy Management*. London, UK: Risk Books.

Trigeorgis, L. 1996. *Real Options: Managerial Flexibility and Strategy in Resource Allocation*. Cambridge, MA: The MIT Press.

Wilmott, P. 2006. *Paul Wilmott on Quantitative Finance*. West Sussex, UK: John Wiley and Sons.

Watkiss, P., A. Hunt, W. Blyth and J. Dyszynski. 2014. The Use of New Economic Decision Support Tools for Adaptation Assessment: A Review of Methods and Applications, Towards Guidance on Applicability. *Climatic Change* 132(3), pp. 401–416.

Winsen, J. K. 2010. An Overview of Project Finance Binomial Loan Valuation. *Review of Financial Economics* 19, 84–89.

Woodward, M., B. Gouldby, Z. Kapelan, S.-T. Khu and I. Townend. 2011. Real Options in Flood Risk Management Decision Making. *Journal of Flood Risk Management* 4, 339–349.

Index

www.ingramcontent.com/pod-product-compliance
Ingram Content Group UK Ltd.
Pitfield, Milton Keynes, MK11 3LW, UK
UKHW020400150926
8624IPUK00004B/263

9 789814 641807